Nucleic Acid Biosensors for Environmental Pollution Monitoring

# Nucleic Acid Biosensors for Environmental Pollution Monitoring

Edited by
**Marco Mascini** and **Ilaria Palchetti**
*Dipartimento di Chimica, Università degli Studi di Firenze, Italy*

RSCPublishing

ISBN: 978-1-84973-131-7

A catalogue record for this book is available from the British Library

© Royal Society of Chemistry 2011

*All rights reserved*

*Apart from fair dealing for the purposes of research for non-commercial purposes or for private study, criticism or review, as permitted under the Copyright, Designs and Patents Act 1988 and the Copyright and Related Rights Regulations 2003, this publication may not be reproduced, stored or transmitted, in any form or by any means, without the prior permission in writing of The Royal Society of Chemistry or the copyright owner, or in the case of reproduction in accordance with the terms of licences issued by the Copyright Licensing Agency in the UK, or in accordance with the terms of the licences issued by the appropriate Reproduction Rights Organization outside the UK. Enquiries concerning reproduction outside the terms stated here should be sent to The Royal Society of Chemistry at the address printed on this page.*

The RSC is not responsible for individual opinions expressed in this work.

Published by The Royal Society of Chemistry,
Thomas Graham House, Science Park, Milton Road,
Cambridge CB4 0WF, UK

Registered Charity Number 207890

For further information see our web site at www.rsc.org

# *Preface*

Nucleic acids are the fundamental bricks of life. They are universal in living things, as they are found in all cells and also in viruses. Their role in nature is of tremendous importance since they are responsible for carrying genetic information from one generation to another.

However, in recent years, the function of nucleic acids has been recognized to extend beyond the Watson–Crick base pair recognition of complementary strands. Their chemical characteristic of being a macromolecule composed of a chain of nucleotides has been used to create unique chemical compounds. Since the early 1990s, many nucleic acid-based molecules, consisting of 40–50 nucleotides—known as *aptamers*—have been isolated from large libraries that are able to bind a broad range of molecules with high affinity and specificity. The molecules and entities that can be recognized by aptamers range from small organic molecules to proteins, cells, and even intact viral particles. A further advance was made in 1994, when for the first time DNA was shown to act as a catalyst. These catalytic DNA molecules are called *DNAzymes*. DNAzymes and aptamers can be combined to form *aptazymes*.

In this book we have attempted to demonstrate the incredible opportunities that nucleic acids can offer to analytical chemistry, and in particular to environmental analytical chemistry. These innovative biomolecules can, in fact, be used to develop smart and innovative biosensors for environmental analysis.

Monitoring of contaminants in the air, water, and soil is instrumental in understanding and managing risks to human health and ecosystems. Given this requirement, as well as the time and cost involved in traditional chemical analysis of environmental samples (*e.g.*, chromatographic methods), there is an expanding need for simple, rapid, cost-effective, and field portable screening methods. A *biosensor* is a compact analytical device incorporating a biological or biologically derived sensing element either integrated within or intimately associated with a physicochemical transducer. The peculiar characteristics of biosensors allow them to complement current environmental field screening and monitoring methods, especially when continuous, real-time, *in situ* monitoring is required. As described in this book, nucleic acids have a pivotal role in development of smart biosensors for environmental monitoring.

Chapter 1 (by Palchetti and Mascini) describes the crucial role of biosensors for environmental monitoring, with special emphasis on the problems involved and the attempts to solve the basic requirements for increased stability, detection sensitivity, and reliability of the biomolecular recognition element. The opportunities that nucleic acids offer in this fields are highlighted. Chapter 2 (by Erdem and Oszoz) provides a description and classification of nucleic acids and their role in living organisms. Chapter 3 (by Palchetti, Marrazza, and Mascini) deals with the use of nucleic acid based biosensors for hybridization and their role in environmental monitoring. Chapter 4 (by Cella, Chen, and Mulchandani) and Chapter 5 (by Nagraj and Lu) focus on the development of biosensors based on aptamers and nucleic acid enzymes respectively for the detection of organic and inorganic pollutants. Chapter 6 (by Rogers and Gary) and Chapter 7 (by Labuda) deal with the use of nucleic acid based biosensors for environmental toxicity screening, and Chapter 8 (by Escosura-Muñiz, Medina, and Merkoçi) focuses on the use of nanomaterials (nanoparticles, quantum dots, nanotubes, *etc.*) as well as miniaturization and lab-on-a-chip technologies for nucleic acid based biosensing systems with interest for environmental applications.

We are fortunate to have assembled contributions from world-class authorities in this field, and we sincerely thank all of them. Sharing their enthusiasm for the field of nucleic acid based biosensors, we have brought their contributions together in this book, which we believe, will be of great help to the increasing number of researchers in this field.

*Marco Mascini and Ilaria Palchetti*

Sesto Fiorentino

# Contents

**Chapter 1  Biosensor Techniques for Environmental Monitoring**    1
*Ilaria Palchetti and Marco Mascini*

    1.1  Introduction: Role of Biosensors in Environmental Analysis    1
    1.2  Biosensors: Definition, Classification and a Brief History    3
    1.3  Innovative Biorecognition Elements for Environmental Analysis    6
    1.4  Conclusions    14
    References    15

**Chapter 2  Nucleic Acids as Biorecognition Element in Biosensor Development**    17
*Arzum Erdem and Mehmet Ozsoz*

    2.1  Description and Classification of Nucleic Acids    17
        2.1.1  Natural Nucleic Acids    17
        2.1.2  Synthetic Nucleic Acids    18
    2.2  Applications of Nucleic Acid Based Biosensor Technologies    21
        2.2.1  DNA Biosensors    25
        2.2.2  RNA Biosensors    27
        2.2.3  PNA Biosensors    28
        2.2.4  LNA Biosensors    29
    2.3  Conclusion    30
    References    30

---

Nucleic Acid Biosensors for Environmental Pollution Monitoring
Edited by Marco Mascini and Ilaria Palchetti
© Royal Society of Chemistry 2011
Published by the Royal Society of Chemistry, www.rsc.org

| | | |
|---|---|---|
| Chapter 3 | **Genosensing Environmental Pollution**<br>*Ilaria Palchetti, Giovanna Marrazza and Marco Mascini* | 34 |

    3.1  Introduction  34
    3.2  Genosensor Development  35
            3.2.1  Probe Design  36
            3.2.2  Probe Immobilization  37
            3.2.3  Sample Treatment and Hybridization  40
            3.2.4  Detection  47
    3.3  Genosensors for Environmental Monitoring  51
    3.4  Conclusions  57
    References  57

| | | |
|---|---|---|
| Chapter 4 | **Aptamer-based Biosensor for Environmental Monitoring**<br>*Lakshmi N. Cella, Wilfred Chen and Ashok Mulchandani* | 61 |

    4.1  Introduction to Aptamers  61
    4.2  Aptamer Properties  62
    4.3  SELEX and its Variants  63
            4.3.1  Bound-state SELEX Variants  64
            4.3.2  Solution-based SELEX Variants  66
            4.3.3  Specialized Variants  66
    4.4  Aptamer-based Biosensors for Environmental
          Monitoring  67
            4.4.1  Detection of Organic and Inorganic Pollutants  68
            4.4.2  Detection of Drugs, EDCs, and PPCPs  70
            4.4.3  Detection of Toxins  72
            4.4.4  Detection of Pathogens  75
            4.4.5  Detection of Nitroaromatic Explosives  77
    4.5  Future Prospects  77
    References  78

| | | |
|---|---|---|
| Chapter 5 | **Catalytic Nucleic Acid Biosensors for Environmental Monitoring**<br>*Nandini Nagraj and Yi Lu* | 82 |

    5.1  Discovery of Catalytic Nucleic Acids  82
    5.2  Detection of Trace Contaminants using Catalytic
          Nucleic Acids as Sensing Platforms  83
    5.3  Isolation of Catalytic Nucleic Acids Using *in vitro*
          Selection  84
    5.4  Conversion of Catalytic Nucleic Acids into Biosensors  85
            5.4.1  Fluorescence Sensing  86
            5.4.2  Colorimetric Sensors  89

Contents ix

|  |  |  | |
|---|---|---|---|
| | | 5.4.3 Dipstick Tests based on AuNP-DNAzyme Conjugates | 91 |
| | | 5.4.4 Electrochemical Sensors | 93 |
| | 5.5 | Expanding the Scope of the Sensing Targets of Catalytic Nucleic Acids by Employing Aptazymes | 93 |
| | 5.6 | Summary and Future Perspectives | 95 |
| | Acknowledgments | | 95 |
| | References | | 95 |

Chapter 6   Nucleic Acid-based Biosensors for the Detection of DNA Damage   99
*Kim R. Rogers and Ronald K. Gary*

- 6.1 Introduction   99
- 6.2 Optical Transduction Schemes   107
- 6.3 Whole-Cell Based Biosensors   110
- 6.4 Field Capability and Stage of Development   116
- 6.5 Future Trends and Summary   117
- Disclaimer   118
- References   118

Chapter 7   Detection of Damage to DNA Using Electrochemical and Piezoelectric DNA-Based Biosensors   121
*Jan Labuda*

- 7.1 Introduction   121
- 7.2 Electrochemical Biosensors   123
  - 7.2.1 Construction of Biosensors for DNA Damage Detection   123
  - 7.2.2 Techniques Used for DNA Damage Detection   124
- 7.3 Piezoelectric Biosensors   133
- 7.4 Conclusions   134
- References   134

Chapter 8   New Trends in DNA Sensors for Environmental Applications: Nanomaterials, Miniaturization, and Lab-on-a-Chip Technology   141
*Alfredo de la Escosura-Muniz, Mariana Medina and Arben Merkoçi*

- 8.1 Introduction   141
- 8.2 Nanomaterial-based Sensors for DNA Detection   142
  - 8.2.1 Optical Sensors   142
  - 8.2.2 Electrochemical Sensors   146
- 8.3 DNA and Nanomaterial-based Sensing Platforms   148

|  |  | 8.3.1 Optical Detection Methods | 149 |
|---|---|---|---|
|  |  | 8.3.2 Electrochemical Detection Methods | 152 |
|  | 8.4 | Nanoprobe- and Nanochannel-based Sensing | 153 |
|  | 8.5 | Lab-on-a-chip Systems | 154 |
|  |  | 8.5.1 Fabrication Technologies | 155 |
|  |  | 8.5.2 Operation | 157 |
|  | 8.6 | Conclusions | 160 |
|  | Acknowledgements | | 161 |
|  | References | | 161 |

**Chapter 9 Conclusions and Criticisms** — **165**
*Ilaria Palchetti and Marco Mascini*

References — 167

**Subject Index** — **168**

CHAPTER 1
# Biosensor Techniques for Environmental Monitoring

ILARIA PALCHETTI AND MARCO MASCINI

Dipartimento di Chimica, Università degli Studi di Firenze, 50019 Sesto Fiorentino (Fi), Italy

## 1.1 Introduction: Role of Biosensors in Environmental Analysis

Monitoring of contaminants in the air, water and soil is an instrumental component in understanding and managing risks to human health and ecosystems. The increasing number of potentially harmful pollutants in the environment calls for fast and cost-effective analytical techniques to be used in extensive monitoring programs. Given this requirement, as well as the time and cost involved in traditional chemical analysis of environmental samples (*e.g.*, chromatographic methods, atomic spectroscopy and related hyphenated techniques), there is an expanding need for innovative methods.

Biosensors appear well suited to complement standard analytical methods for a number of environmental monitoring applications.[1,2] The main advantages offered by biosensor technology over conventional analytical techniques are fast and economical measurements; the possibility of miniaturization and portability; the possibility of continuous monitoring; and, in some cases, the ability to measure pollutants in complex matrices with minimal sample preparation. Although many of the systems that have been developed cannot compete with conventional analytical methods in terms of accuracy and reproducibility, they can be used by regulatory authorities and by industry to provide enough information for routine testing and screening of samples. The peculiar characteristics of biosensors allow these devices to complement current field screening and monitoring methods such as the ELISA test, especially when continuous, real-time, *in situ* monitoring is required.

Some excellent reviews have summarized the recent progress in use of biosensors for environmental applications.[1–9] Biosensors have been used in the analysis of chemical compounds, such as pesticides, heavy metals, endocrine disruptors (*e.g.*, phenol derivatives) and persistent organic pollutants (*e.g.*, PCB), as well as for the evaluation of environmental quality parameters (related to chemical pollution) such as biological oxygen demand (BOD). Moreover, biosensors have been also used for the monitoring of microbial pollution and detection of environmentally relevant organisms—microorganisms and plant or animal species. In other words, they can be used to control chemical and microbiological contamination and pollution as well as to check ecosystem biodiversity. Moreover, in recent years, the use of biosensors has been successfully proposed in the areas of environmental toxicity, cytotoxicity and genotoxicity. All these applications lead to the next challenge for environmental biosensor applications—the assessment of environmental pollution exposure and its impact on fundamental biological processes. Obviously, this "will require sophisticated computer models that can handle the immense volume and complexity of data generated for each individual and, also, would allow for integration of data on environmental exposures with genetic factors for the individual and the population".[10]

Recently, biosensor technology has benefited from results obtained in other fields, such as biotechnology, nanoengineering and nanotechnology. "Lab on a chip" technology has greatly increased the possibility of obtaining easy-to-use and self-contained devices, minimizing the need for sample processing and analysis in the laboratory. Developments in micro- and nanotechnology continually help to miniaturize the devices. Biotechnologies help to increase the number of stable, sensitive and selective sensing elements that can be used for obtaining reliable biosensors.

Nevertheless, the commercial exploitation of biosensors for environmental applications is still at an early stage (with some exceptions). From a technical point of view this is still predominately tied to the stability, detection sensitivity and reliability of the biomolecular recognition element. In addition (as clearly stated in ref. 11), before a biosensor gains market acceptance, it must prove capable of being validated by well-established procedures. Research studies, as reported in literature, with few real samples or treated samples, often fail to provide an adequate measure of capability for "real-world" samples, leading to failed technology transfer and further investment. Such activities require appropriate sources of finance for technology development and demonstration, and undoubtedly these economic resources are more concentrated in the clinical and medical fields than in environmental science. Finally, the success of biosensors must prove that they are the inevitable choice as a cost-effective analytical tool.

Bearing these facts in mind, in this chapter we aim to give an overview of the biosensing elements used for environmental applications, focusing on their innovative aspects. Descriptions of the technical attempts to increase stability and sensitivity of the (bio)receptors are reported. Particular emphasis has been given to their use in environmental analysis.

## 1.2 Biosensors: Definition, Classification and a Brief History

The history of biosensors starts around 60 years ago. In 1956 Professor Leland C. Clark published his paper on the development of an oxygen probe; from this initial research activity he expanded the range of analytes that could be measured, and in 1962, in a conference at the New York Academy of Sciences, he described how "to make electrochemical sensors (pH, polarographic, potentiometric or conductometric) more intelligent" by adding "enzyme transducers as membrane enclosed sandwiches".[12] The first example was illustrated by entrapping the enzyme glucose oxidase in a dialysis membrane over an oxygen probe. The addition of glucose proportionally determined the decrease of oxygen concentration. This first biosensor was described as an *enzyme electrode*.[13] Then subsequently, in 1967, Updike and Hicks used the same term to describe a similar device where again glucose oxidase was immobilized in a polyacrylamide gel onto the surface of an oxygen electrode for the rapid and quantitative determination of glucose.[14] In addition to amperometry, Guilbault and Montalvo in 1969 used glass electrodes coupled with urease to measure urea concentration by potentiometric measurement.[15]

Since then there has been a rapid proliferation of biosensors with different biological elements and different transducers. This intensive research effort has led to the commercial exploitation of some devices. The first biosensor that appeared on the market was the glucose biosensor for diabete care. In 1984 a paper by Cass *et al.* described the use of ferrocene and its derivatives as mediators for amperometric biosensors.[16] A few years later the Medisense Exatech (now Abbott) Glucose Meter was launched in the market and become the world's best-selling biosensor product. The initial product was a pen-shaped meter with disposable screen-printed electrodes; nowadays many formats are commercially available from different companies worldwide. During the same years (mid 1980s), researchers from Pharmacia started to work jointly with physics and biochemistry faculty at Linkoping University in Sweden, in order to develop a new bioanalytical instrument that could monitor the interactions between biomolecules. In 1984 a new company, Pharmacia Biosensor, was created. This company introduced a new instrument, the BIAcore, in 1990 (it is nowadays distributed by General Electric).

The concept of a biosensor has evolved enormously during the years. In the early days of this research area, many authors considered a biosensor to be a self-contained analytical device that responded to the concentration of chemical species in biological samples, without mentioning the role of the biologically active material involved in the device. This obviously caused many misunderstandings, since any physical or chemical sensor operating in biological samples could be considered a biosensor.[11] Thus, many authors started to reserve the use of the term "biosensor" for a chemical sensor in which the recognition system utilizes a biochemical mechanism.[17,18]

In early 2000, two Divisions of the International Union of Pure and Applied Chemistry (IUPAC), namely Physical Chemistry (Commission I.7 on Biophysical Chemistry, formerly Steering Committee on Biophysical Chemistry) and Analytical Chemistry (Commission V.5 on Electroanalytical Chemistry) prepared recommendations on the definition, classification and nomenclature relating to electrochemical biosensors; these recommendations have been then extended to other types of biosensors.[19] Following these IUPAC recommendations,

*a biosensor is defined as a self-contained integrated device, which is capable of providing specific quantitative or semi-quantitative analytical information using a biological recognition element (biochemical receptor) which is retained in direct spatial contact with an electrochemical transduction element. Because of their ability to be repeatedly calibrated, a biosensor should be clearly distinguished from a bioanalytical system, which requires additional processing steps, such as reagent addition. A device that is both disposable after one measurement, i.e., single use, and unable to monitor the analyte concentration continuously or after rapid and reproducible regeneration, should be designated a single use biosensor.*[19]

Another important concept was introduced, some years later, by Turner and Newman;[20] they referred to a biosensor as "a compact analytical device incorporating a biological or biologically-derived sensing element either integrated within or intimately associated with a physicochemical transducer", thus including synthetic chemical compounds that mimic the biological material in the development of biosensors. Nowadays, the concept of "biologically derived element" is fully accepted in the scientific community.

Biosensors are classified according to the biological specificity-conferring mechanism (Figure 1.1) or, alternatively, to the mode of physicochemical signal transduction. They may be further classified according to the analytes or reactions that they monitor; for example, direct monitoring of analyte concentration or of reactions producing or consuming such analytes. Alternatively, the indirect monitoring of inhibitor or activator of the biological recognition element (biochemical receptor) may be used.

In terms of transduction principles, biosensors can be classified as optical, electrochemical, mass, magnetic, calorimetric or micromechanical biosensors.

- *Optical detection* by fluorescence spectroscopy is a popular method, largely because of the ease with which biomolecules (especially nucleic

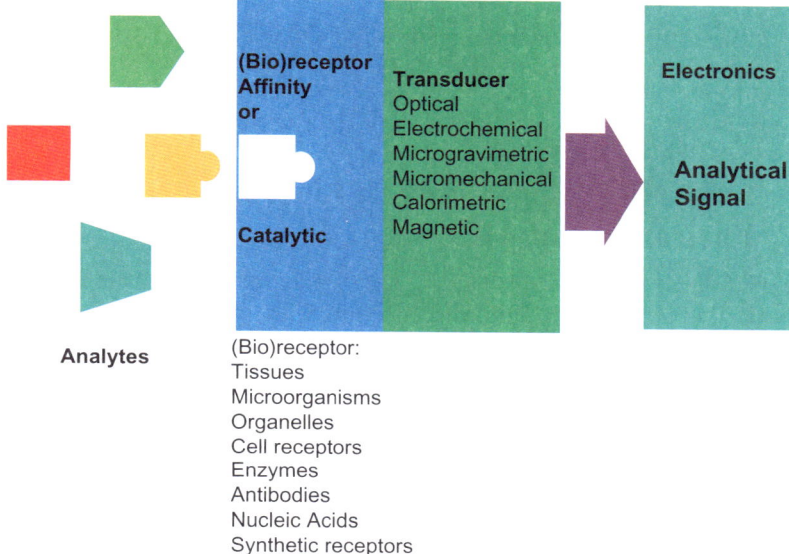

**Figure 1.1** Schematic representation of a biosensor.

acids) can be fluorescently labeled, the availability of many different fluorophores and quenchers, and the inherent capability for real-time multiplex detection.[22] Chemiluminescence is another widely used optical technique. A different type of optical transduction, based on an evanescent wave device, can offer real-time label-free optical detection. These biosensors rely on monitoring changes in surface optical properties (shift in resonance angle due to change in the interfacial refractive index) resulting from the surface binding reaction.

- *Electrochemical devices* have also proved very useful, because of their inherent miniaturization and their compatibility with advanced microfabrication technology. Electrochemical detection usually involves monitoring a current response under controlled potential conditions. However, other changes in electrochemical parameters such as capacitance, impedance and conductivity have been used.
- Another useful label-free, *mass detection* scheme relies on the use of quartz crystal microbalance (QCM) transducers. A QCM biosensor consists of an oscillating crystal with a bioreceptor immobilized on its surface. The increased mass, associated with the biorecognition reaction, results in a decrease of the oscillating frequency. Acoustic wave sensors used in thickness-shear mode with a liquid sample detect changes in a number of physical properties including mass, viscosity and charge density.
- *Micromechanical* transduction based on cantilevers and label-free biosensors capable of detecting biomolecular interactions via the bending of microfabricated cantilevers coated with bioreceptors have been reported in the literature.

- Finally, in *magneto-biosensors*, magnetic labels are used to detect magnetoresistance, giant magnetoresistive effect (GMR), spin-value GMR and other parameters.

## 1.3 Innovative Biorecognition Elements for Environmental Analysis

"The biological recognition element may be based on a chemical reaction catalysed by, or on an equilibrium reaction with macromolecules that have been chemically synthesized, naturally isolated, or engineered, or are present in their original biological environment. In the latter cases, equilibrium is generally reached and there is little or no further net consumption of analyte(s) by the immobilized biocomplexing agent incorporated into the sensor".[19]

Enzymes (and all biological elements, such as tissues, cells, microorganisms, which contain enzymes) represent the class of what are now called *catalytic elements*. Enzymes were historically the first molecular recognition elements included in biosensors, and continue to be the basis for a significant number of publications reported for biosensors in general as well as for environmental applications. Enzyme biosensors have several advantages. These include a stable source of material (primarily through biorenewable sources); the possibility of modifying the catalytic properties or substrate specificity by means of genetic engineering; and catalytic amplification of the biosensor response by modulation of the enzyme activity with respect to the target analyte. However, enzyme-based biosensors show also some limitations for use in environmental applications. These include the limited number of substrates for which enzymes have been evolved, the limited interaction between environmental pollutants and specific enzymes, and in the case of inhibitor formats, the lack of specificity in differentiating among compounds of similar classes such as nerve agents or organophosphate (OP) and carbamate pesticides.

Typical examples of enzymes involved in environmental applications are cholinesterase for OP and carbammate pesticide analysis, and tyrosinase for analysis of phenols and related compounds with endocrine disruptor characteristics. As already mentioned, genetic engineering helps in the careful selection of the location and type of mutations giving rise to enzymes with enhanced or particular properties, such as higher affinity towards specific analytes, higher stability, higher electron transfer rates, and residues able to provide an oriented or more stable immobilization. These improvements have already resulted in biosensors with enhanced performance.[3] Recent progress with respect to genetically modified enzyme biosensors for environmental applications has been well reviewed.[1,3-5]

An emerging field of biorecognition elements is so-called *whole-cell systems*. Whole cells have long been used for environmental applications, in particular for BOD monitoring.[21] However, recently they have benefited enormously from the recent improvements in recombinant DNA technology[7] and there is renewed interest in their use in monitoring environmental pollution and toxicity. Whole-cell systems are based on complex cellular functions, among which enzyme-catalysed reactions play an important role. The biosensors are constructed by the fusion of promoters (responsive to the relevant environmental conditions) to easily monitored reporter genes. Depending on the choice of reporter gene, expression can be monitored by the production of colour, light, fluorescence or electrochemical reactions. Although there are numerous examples of genetic modification involving bacteria, yeast, algae and tissue culture cells, *genetically engineered bacteria* (GEMs) are most often reported in cell-based biosensors.

Antibody-based biosensors (*immunosensors*) are inherently more versatile than enzyme-based biosensors in that antibodies have been generated which specifically bind to individual compounds or groups of structurally related compounds with a wide range of affinities. Antibodies are the most used affinity proteins in all life science applications. They can be isolated to many targets with high affinity and specificity. The most used antibody type, the IgG molecule (Figure 1.2), is a 150-kDa protein composed of four polypeptide

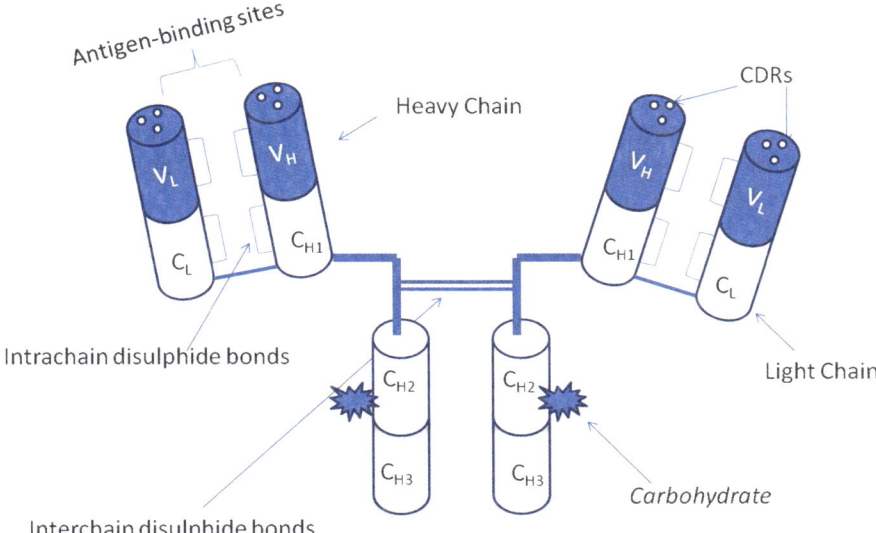

**Figure 1.2** Diagrammatic representation of an immunoglobulin G (IgG) molecule. The IgG molecule is composed of two identical light chains and two identical heavy chains. The light chains are composed of a variable (VL) and constant (CL) domain. The heavy chain consists of one variable (VH) and three constant (CH1, CH2 and CH3) domains with a hinge region connecting the CH1 and CH2 regions. The heavy and light chains are connected via disulfide bonds; disulfide bonds are also present in the constant and variable regions. The complementary-determining regions (CDRs) at the amino-terminal of the variable domains confer antigenic specificity and contain considerable amino acid sequence variation. Adapted from ref. 22.

chains, two identical larger heavy chains and two identical shorter light chains. Each light chain is coupled to a heavy chain via a disulfide bond and the two heavy–light dimers are correspondingly disulfide bridged, forming the typical Y-shaped antibody structure. In early experimental observations antibodies were digested with papain protease, generating two identical antigen-binding fragments (Fab, fragment antigen binding) and one without antigen binding activity (Fc, fragment crystallizable) and these terms are still used to describe the antibody structure.

The stability of IgG is strictly dependent on disulfide bonds; moreover, this large, bivalent, multidomain protein possesses a complex glycosylation pattern. These characteristics lead to relatively poor heat stability and a comparatively difficult and expensive manufacturing process. In addition, antibodies use only a minor part of the molecule for antigen recognition. Large domains have structural function, and there are other defined binding sites that are responsible for interaction with complement factors and various Fc receptors; these interactions may be desired or even essential for some applications, but increase the complexity of the molecules.

The main antibody formats current available for use in immunosensors are polyclonal, monoclonal and recombinant antibodies (Figure 1.3).[22]

- Antibodies can be produced by immunizing animals with an antigen. The animal will then generate a pool of antibodies towards different epitopes of the antigen. These antibodies produced from immunization are referred to as *polyclonal antibodies*. They have different amino acid sequences recognizing different epitopes on the same antigen, and are polyspecific.
- *Monoclonal antibodies*, *i.e.*, antibodies directed to a single epitope, were first introduced by Köhler and Milstein in 1975 through their hybridoma technology.[23] This technology earned Köhler and Milstein the Nobel Prize in 1984. In hybridoma technology an antibody-producing B-cell is fused with a myeloma cancer cell. The generated hybrid cell inherits the capacity for antibody production from the B-cell and immortal growth from the tumor cell, providing the possibility of indefinite production of a specific antibody. Monoclonal antibodies are potentially monospecific and are generated by immortalized cell lines (hybridomas) rendering them more homogeneous than immune serum (containing antibodies obtained from an immunized animal). However, the laborious nature of monoclonal antibody generation, coupled with their mainly murine host dependency (the antibodies produced are of rodent origin, most commonly mouse), has led to the investigation of recombinant antibodies to produce optimized biorecognition elements.
- *Recombinant antibodies* are encoded, selected, and expressed in multiple structural formats, which can have considerably different biophysical attributes. The simplest and commonly used structural format is the single chain fragment (scFv) in which the V-regions of antibodies are linked by a flexible peptide linker. The most common linkers are based on

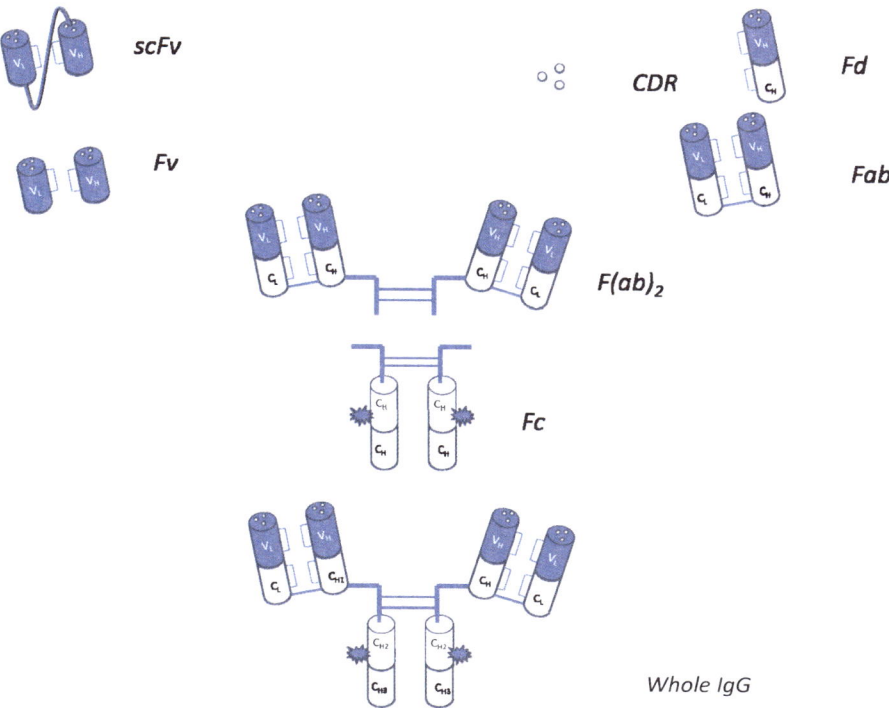

**Figure 1.3** Diagrammatic illustration of antibody and antibody fragments which can be produced by genetic, chemical, or enzymatic means. The antibody may be broken up into either Fab (single antigen-binding fragment) or F(ab)2 (two antigen-binding fragments), and Fc (crystalline fragments) regions. Fab fragments may be further broken up into Fv (variable fragment) and scFv (variable fragment stabilized with a synthetic linker) and CDR regions, which are the smallest fragment capable of antigen binding. Adapted from ref. 22.

glycine–serine repeat structures and can be of different lengths, depending on the intended valency of the molecule. Another commonly used recombinant antibody format is the Fab (fragment antigen binding) molecule.

*Phage display* has proved an useful tool for the isolation of recombinant antibody fragments with desired specificities. The technique involves the display of a library of single-chain antibody (scFv) fragments on the surface of filamentous phages, followed by selection of the desired recombinant phages by means of specific binding to an antigen of interest. Although phage display has advantages over conventional polyclonal and monoclonal antibody production, few authors have employed this technique for the selection of reagents for the detection of environmental pollutants, either chemical or microbial. Phage display also provides several approaches for rational improvement of antibody affinity and selectivity. Given a set of phage clones with known affinity and selectivity profiles, selection strategies can be designed to isolate clones with optimal properties from the existing library of clones. By correlating affinity and selectivity data with DNA sequence information, it is feasible to design and construct novel sequences that express antibodies with the desired properties.

Today, a number of other techniques exist for generation of recombinant antibodies, including yeast display, bacterial surface display and ribosome display. All, apart from ribosome display, are cell-based methods, which apply the same basic fundamental principles as phage display. *Ribosome display* is a completely *in vitro* system that relies on the translation of artificially produced antibody mRNA (usually of extremely high diversity), which encodes a linker peptide after the antibody fragment, but does not contain a stop codon. Therefore, on translation, the protein is produced but remains linked to the ribosome and the encoding RNA as a large complex. Specific antibodies can then be isolated using a typical biopanning method, with retrieval of binding clones by Reverse Transcriptase PCR (RT-PCR).

Although significant progress has been made in selecting stable recombinant antibody fragment libraries, viable alternative methods are still required. Thus, the increasing experience in the field of combinatorial libraries and protein engineering has inspired researchers to develop new "nonimmunoglobulin affinity" protein without the limitations of antibodies. Consequently, antibodies are today facing increasing competition from a large number of so-called *engineered protein scaffolds*. However, these new scaffold proteins are mainly applied to clinical targets, and to our knowledge no example of an environmental application has yet been reported in literature.

Another class of interesting biosensing molecules is *nucleic acids* (NA). In recent years, NA have been incorporated into a wide range of biosensors and bioanalytical assays, because of their wide range of physical, chemical and biological activities.[1,24–28] As is generally known, NA molecules have the function of carrying and passing on genetic information. From an analytical point of view this can be exploited for the specific identification of animal and plant species, genetically modified organisms, bacteria, viruses, toxins, *etc*. Genosensors, in particular, result from the integration of a sequence-specific probe (usually a short synthetic oligonucleotide) and a signal transducer. The probe, immobilized onto the transducer surface, acts as the biorecognition molecule and recognizes the target DNA or RNA via hybridization reaction.

However, about 20 years ago, NA began to find a new role in the field of materials science and biotechnology.[26–28] *Aptamers*, for instance, are single-stranded DNA or RNA ligands which can be selected for different targets, starting from a huge library of molecules containing randomly created sequences.[28] The selection process is called *systematic evolution of ligands by exponential enrichment (SELEX)*, first reported in 1990.[29–31] The SELEX process involves iterative cycles of selection and amplification starting from a large library of oligonucleotides with different sequences (generally $10^{15}$ different structures). After the incubation with the specific target and the partitioning of the binding from the nonbinding molecules, the oligonucleotides that are selected are amplified to create a new mixture enriched in those nucleic acid molecules having a higher affinity for the target. After several cycles of the selection process, the pool is enriched in the high-affinity sequences at the expense of the low-affinity binders. The number of cycles required depends on the stringency conditions, but, once obtained and once the sequence is known, unlimited amounts of the aptamer can be easily achieved by chemical synthesis.

In addition to this very important aspect of having an unlimited source of identical affinity recognition molecules, aptamers can offer advantages over antibodies that make them very promising for analytical applications. The main advantage is avoiding the use of animals or cell lines for the production of the molecules. Antibodies against molecules that are not immunogenic, for instance, are difficult to generate. On the contrary, aptamers are isolated by *in vitro* methods that are independent of animals and an *in vitro* combinatorial library can be generated against any target. In addition, generation of antibodies *in vivo* means that is the animal immune system that selects the sites on the target protein to which the antibodies bind. The *in vivo* parameters restrict the identification of antibodies that can recognize targets only under physiological conditions, limiting the extension to which the antibodies can be functionalized and applied. In contrast, the aptamer selection process can be manipulated to obtain aptamers that bind a specific region of the target and with specific binding properties in different binding conditions.

After selection, aptamers are produced by chemical synthesis and purified to a very high degree by eliminating the batch-to-batch variation found when using antibodies. By chemical synthesis, modifications in the aptamer can be introduced enhancing the stability, affinity and specificity of the molecules. Often the kinetic parameters of aptamer–target complex can be changed for higher affinity or specificity. Another advantage over antibodies is the higher temperature stability of aptamers and the fact that they can recover their native active conformation after denaturation, whereas antibodies are large, temperature-sensitive proteins that can undergo irreversible denaturation.

Moreover, in recent years NA, particularly DNA, have also been used to create systems capable of catalytic activity. (DNA molecules are less susceptible to hydrolysis than RNA, and thus are highly stable.) The term *nucleic acid enzyme* is used to identify these NA structures that have catalytic activity. Ribozymes (literally enzymes made of RNA) are found in nature and mediate phosphodiester bond cleavage and formation and peptide bond formation. In 1989, Thomas R. Cech and Sidney Altman won the Nobel Prize in chemistry for their "discovery of catalytic properties of RNA". Artificial ribozymes have been obtained by means of combinatorial chemistry approaches, such as *in vitro* selection and *in vitro* evolution,[32] and have been shown to catalyze quite a broad array of other chemical reactions.[32]

*DNAzymes* are DNA-based biocatalysts capable of performing chemical transformations.[33] They are artificial molecules, not found in nature; all known DNAzymes have been isolated by *in vitro* selection. So far, most of their substrates have been found to be nucleic acids. DNAzymes can therefore provide additional control over NA-based devices and, because DNAzymes often catalyze multiple turnover reactions, such devices can have amplification effects. Among the many classes of DNAzymes, RNA-cleaving DNAzymes are the most widely used, mainly because of their simple reaction conditions, fast turnover rates and significant modifications of their substrate lengths.[26]

DNAzymes can perform chemical modifications on nucleic acids, while aptamers can bind a broad range of molecules. A combination of the two has generated a new class of functional nucleic acids known as allosteric DNAzymes or *aptazymes*.

Furthermore, even genomic DNA has been employed to develop biosensors, and in particular biosensors for toxicity evaluation. The elaboration of new, highly sensitive, specific and rapid methods of determination of environmentally toxic compounds has always been a challenge to analytical chemists. This is due to the complexity of the ecological situation and the demand for diagnosing the consequences of the impact of biologically active contaminants on the human organism. For these reasons, the development and characterization of rapid, sensitive and inexpensive assays for detection of chemically induced damage to DNA caused by pollutants is an important issue in environmental monitoring. A significant number of short-term tests for genotoxicity/mutagenicity have been developed to determine the extent of environmental hazards in polluted water and sediments. These bioassays have been used in well-constructed batteries of cytotoxicity and genotoxicity tests to complement traditional approaches. These tests typically fall into one of several classes, depending on their mechanism and end-point. These classes include bacterial mutagenesis, cultured mammalian cell mutagenesis, chromosomal damage and DNA damage. Despite their description as "short term", many of these assays are expensive to run, require sophisticated technical expertise and are not well suited to be adapted as screening applications. In recent years, there has been an enormous increase in the use of nucleic acids as a tool in the recognition or monitoring of chemical compounds of environmental interest that may affect the genome. In this context, DNA-based biosensors can be a valid approach to the detection of DNA damage and can be designed to be rapid, inexpensive and user friendly. Different kind of transducers have been employed, each of them having their own advantages and disadvantages.

In summary, as reported in previous sections, NA biosensing is an expanding field. Genosensors offer considerable promise for obtaining sequence-specific information in a faster, simpler and cheaper manner than traditional hybridization assays. Many examples of DNA biosensors for the detection of DNA damage and interaction have also been reported. Moreover, although their appearance in the literature has been very recent, functional DNA molecules such as aptamers, DNAzymes and aptazymes have already found application in almost every aspect of DNA nanotechnology, and the resulting new materials and devices may penetrate into many other fields for practical application, including environmental monitoring.

An interesting class of synthetic molecules for biosensor development is the *molecularly imprinted polymers (MIPs)*. MIPs are synthetic polymeric materials with specific recognition sites complementary in shape, size and functional groups to the template molecule, involving an interaction mechanism based on molecular recognition. These recognition sites mimic the binding sites of biological entities such as antibodies and enzymes (Figure 1.4). In particular, in the production of Mips, functional and cross-linking monomers are copolymerized in the presence of a target analyte (the imprint molecule), which acts as a molecular template. The functional monomers initially form a complex with the imprint molecule; following polymerization, their functional groups are held in position by the highly cross-linked polymeric structure. Subsequent removal of the imprint molecule reveals binding sites that are complementary in size and shape to the analyte. In this way, something like a molecular memory is introduced into the polymer, which is now capable of selectively rebinding the analyte.[34] Mips are interesting because of their stability, ease of preparation and low cost for most of the target analytes for numerous applications. They can be prepared by a number of methods that differ in the way the template is linked to the functional monomer and subsequently to the polymeric binding sites. Thus, the template can be linked and subsequently recognized by reversible covalent bonds, metal ion coordination or noncovalent bonds. Despite the interest in the covalent approach, which gives rise to higher stability of the prepolymerization complex, the most widely applied technique to generate molecular binding sites is based on noncovalent self-assembly of the template

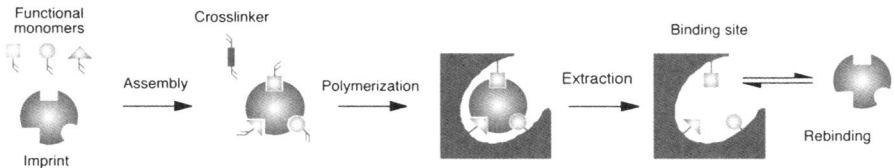

**Figure 1.4** Schematic representation of the molecular imprinting principle. Functional and cross-linking monomers are copolymerized in the presence of a target analyte (the imprint molecule), which acts as a molecular template. The functional monomers initially form a complex with the imprint molecule, and following polymerization, their functional groups are held in position by the highly cross-linked polymeric structure. Subsequent removal of the imprint molecule reveals binding sites that are complementary in size and shape to the analyte. In that way, a molecular memory is introduced into the polymer, which is now capable of selectively rebinding the analyte. Reproduced from ref. 34 by permission of The Royal Society of Chemistry.

with functional monomers prior to polymerization with cross-linking monomer, and then template extraction followed by rebinding via noncovalent interactions.[35,36]

The potential of MIPs has been so widely demonstrated that several companies are now selling imprinted polymers such as Biotage (Mips as phase for solid-phase extraction in Affinilute SPE, www.biotage.com), PolyIntell (www.polyintell.com), Semorex (www.semorex.com) and Supelco (SupelMip, www.sigma-aldrich.com/supelmip). Although most of the development of MIPs has been carried out in the biological and clinical fields, their potential as selective tools in analytical techniques for environmental applications has been well illustrated in numerous publications. *Solid-phase extraction (SPE)* is certainly the most active and advanced area, as highlighted by numerous relevant applications of MIPs for the selective extraction of target analytes in real complex matrices.

However, MIPs are not intrinsically selective. Their selectivity results from the combination of a polymerization procedure that produces specific cavities for the target analytes, together with an associated usage protocol that should favour the development of specific interactions. For the use of MIPs in separation or in SPE, selectivity is controlled by the choice of the solvent required to favor these interactions. For pseudo-immunoassay and sensors, their direct application to real matrices is more difficult because the conditions of use are mainly fixed by the nature of the sample; direct and selective binding of the target molecule in aqueous media is still difficult in most cases.

## 1.4 Conclusions

Sensor technologies hold exceptional promise for providing critical information for continuous, real-time and *in situ* data collection. Simultaneous measurement of multiple agents (*multiplexing*) within a single device has been also demonstrated. New sensing modalities have emerged from biotechnology, nanotechnology and nanoengineering that could be adapted and developed for environmental monitoring.

In addition to being self-contained, biosensors are capable of quantitative, continuous data capture in the field, without the need for sample processing and analysis in a laboratory. However, to be competitive with other existing technologies, these devices must be easy to use, portable, minimally inconvenient, rugged and inexpensive to deploy.

In our opinion, all these aspects are within the reach of existing manufacturing technologies and immediate development. The real challenge, nowadays, is the possibility of obtaining stable (bio)sensing molecules.

As reported in this chapter, artificial receptors such as aptamer, aptamzymes, DNAzymes, MIPs and scaffold affinity proteins have potential as stable surrogates for biological recognition agents such as antibodies, enzymes, tissues, or cells, increasing stability and versatility of the final biosensor device. In particular, NAs, with their inherent chemical characteristics, have greatly increased their potential application in many areas of science, including environmental analysis. Undoubtedly NAs should have a key role in the identification of new bioreceptors for the development of smart biosensors.

## References

1. K. R. Rogers, *Anal. Chim. Acta*, 2006, **568**, 222.
2. S. Rodriguez-Mozaz, M. J. Lopez de Alda and D. Barceló, *Anal. Bioanal. Chem.*, 2006, **386**, 1025.
3. M. Campàs, B. Prieto-Simónc and J. L. Marty, *Semin. Cell. Dev. Biol.*, 2009, **20**, 3.
4. M. Tudorache and C. Bala, *Anal. Bioanal. Chem.*, 2007, **388**, 565.
5. A. Amine, H. Mohammadi, I. Bourais and G. Palleschi, *Biosens. Bioelectron.*, 2006, **21**, 1405.
6. S. Rodriguez-Mozaz, M. J. Lopez de Alda, M. P. Marco and D. Barcelo, *Talanta*, 2005, **65**, 291.
7. E. Z. Ron, *Curr. Opin. Biotechnol.*, 2007, **18**, 252.
8. S. Rodriguez-Mozaz, M. J. Lopez de Alda and D. Barcelo, *J. Chromatogr. A.*, 2007, **1152**, 97.
9. S. Andreescu and O. A. Sadik, *Pure Appl. Chem.*, 2004, **76**, 861.
10. D. Schwartz and F. Collins, *Science*, 2007, **316**, 695.
11. J. H. T. Luong, K. B. Male and J. D. Glennon, *Biotechnol. Adv.*, 2008, **26**, 492.
12. L. C. Clark, *Trans. Am. Soc. Artif. Intern. Organs*, 1956, **2**, 41.
13. L. C. Clark and C. Lyons, *Ann. N. Y. Acad. Sci.*, 1962, **102**, 29.
14. S. J. Updike and G. P. Hicks, *Nature*, 1967, **214**, 986.
15. G. G. Guilbault and J. Montalvo, *J. Am. Chem. Soc.*, 1969, **91**, 2164.
16. A. E. G. Cass, D. G. Francis, H. A. O. Hill, W. J. Aston, I. J. Higgins, E. V. Plotkin, L. D. L. Scott and A. P. F. Turner, *Anal. Chem.*, 1984, **56**, 667.
17. K. Cammann, *Fresenius Z. Anal. Chem.*, 1977, **287**, 1.
18. A. P. F. Turner, I. Karube and G. S. Wilson, *Biosensors, Fundamentals and Applications*, Oxford University Press, Oxford, 1987.
19. D. R. Thevenot, K. Toth, R. A. Durst and G. S. Wilson, *Biosens. Bioelectron.*, 2001, **16**, 121.
20. J. D. Newman, P. J. Warner, A. P. F. Turner and L. J. Tigwell, *Biosensors: A Clearer View*, Cranfield University, Cranfield, Beds, UK, 2004.
21. H. Nakamura, *Anal. Methods*, 2010, **2**, 430.
22. R. O. Kennedy, W. J. J. Finlay, P. Leonard, S. Hearty, J. Brennan, S. Stapleton, S. Townsend, A. Darmaninsheehan, A. Baxter and C. Jones, in *Sensors for Chemical and Biological Applications*, ed. M. K. Ram and V. R. Bhethanabotla, CRC Press, Boca Raton FL, 2010, p. 195.
23. G. Köhler and C. Milstein, *Nature*, 1975, **7**, 495.
24. G. Bagni, D. Osella, E. Sturchio and M. Mascini, *Anal. Chim. Acta*, 2006, **573–574**, 81.

25. J. Wang, G. Rivas, X. Cai, E. Palecek, P. Nielsen, H. Shiraish, N. Dontha, D. Luo, C. Parrado, M. Chicharro, P. A. M. Farias, E. S. Valera, D. H. Grant, M. Ozsoz and M. N. Flair, *Anal. Chim. Acta*, 1997, **347**, 1.
26. Y. Lu and J. Liu, *Curr. Opin. Biotechnol.*, 2006, **17**, 580.
27. Y. Ito and E. Fukusaki, *J. Mol. Catal. B: Enzym.*, 2004, **28**, 155.
28. S. Tombelli, M. Minunni and M. Mascini, *Biomol. Eng.*, 2007, **24**, 191.
29. A. D. Ellington and J. W. Szostak, *Nature*, 1990, **346**, 818.
30. C. Tuerk and L. Gold, *Science*, 1990, **249**, 505.
31. R. R. Breaker and G. F. Joyce, *Chem. Biol.*, 1994, **1**, 223.
32. R. Fiammengo and A. Jaschke, *Curr. Opin. Biotechnol.*, 2005, **16**, 614.
33. N. K. Navani and Y. Li, *Curr. Opin. Chem. Biol.*, 2006, **10**, 272.
34. K. Haupt, *Analyst*, 2001, **126**, 747.
35. L. Ye and K. Haupt, *Anal. Bioanal. Chem.*, 2004, **378**, 1887.
36. V. Pichon and F. Chapuis-Hugon, *Anal. Chim. Acta.*, 2008, **622**, 48.

CHAPTER 2
# Nucleic Acids as Biorecognition Element in Biosensor Development

ARZUM ERDEM AND MEHMET OZSOZ

Ege University, Faculty of Pharmacy, Analytical Chemistry Department, 35100 Bornova, Izmir, Turkey

## 2.1 Description and Classification of Nucleic Acids

### 2.1.1 Natural Nucleic Acids

Nucleic acids (NAs) are basic hereditary molecules of the living organism and store, transport and translate the genetic information. The NAs lack sulfur; unlike lipids and proteins, they are phosphate-rich compounds. NAs are huge linear polymers which can exist as both natural and artificial structures. DNA (deoxyribonucleic acid) and RNA (ribonucleic acid) are the most important NAs. DNA and RNA are chemically similar. The primary structures of both are linear polymers composed of monomers called *nucleotides*. The nucleotides are composed of a nitrogenous base, a pentose group and a phosphate molecule. These nucleotides are linked together with 5′–3′ phosphodiester bonds. The organic base can be either a purine (adenine or guanine) or a pyrimidine (thymine, cytosine or uracil). The genetic information is needed to maintain a living organism, and is carried by genes which are made up of DNA sequences. Passage of the genetic information from a cell to its daughter cell is done by complementary base pairing. Therefore, the most important biological function of DNA is to carry hereditary knowledge. Additionally, via translation, the DNA sequence is matched with its complementary sequence in RNA which is further used to make a protein that satisfies the translated code.[1]

### 2.1.1.1 DNA

DNA is a double helix and has two complementary antiparallel strands, one running in the 5' to 3' direction and the other from 3' to 5' (structural formula 2.1). The nucleotides in the each strand are connected to each other by ester bonds. The pentose sugar in DNA is deoxyribose, which does not have oxygen in its C-2' position, unlike ribose.

DNA consists of the pyrimidine bases adenine (A) and guanine (G) and purine bases thymine (T) and cytosine (C). Each A in one strand makes two hydrogen bonds with T in the other strand, and each G in one strand makes three hydrogen bonds with C in the other strand. The nitrogenous base molecules are weakly basic compounds and this is why they are called bases.[2]

### 2.1.1.2 RNA

DNA is the principal molecule for genetic information. However, there is another biologically important macromolecule that consists of a chain of nucleotide units: RNA. RNA usually exists as single strand inside the cell, whereas DNA is double stranded in its original form. As can be seen from structural formula 2.2, the structure of RNA is essentially the same as that of DNA, except for some minor differences such as the replacement of thymine (T) by uracil (U), which lacks a methyl group in the C-5' position on its purine ring. Another difference is that the RNA chain has a ribose sugar in its backbone. Ribose is also a pentose sugar but, unlike deoxyribose, it possesses oxygen in its C-2' position. The phosphate backbone of the molecule is the same as the DNA molecule.There are three main types of RNA molecule present in a cell: *messenger RNA (mRNA)*, *transfer RNA (tRNA)* and *ribosomal RNA (rRNA)*.[2]

MicroRNAs *(miRNAs)* are noncoding RNA sequences approximately 22 nucleotides in length. Their biogenesis is peculiar, they do not produce a functional protein, and their specific biological function is still unclear.[3,4]

## 2.1.2 Synthetic Nucleic Acids

### 2.1.2.1 Peptide Nucleic Acid

Peptide nucleic acid (PNA) is the synthetic analog of natural NA in which the sugar phosphate backbone of natural NA has been replaced by a synthetic peptide backbone. As can be seen from structural formula 2.3, the repeating

Structural formula 2.1: Deoxyribonucleic acid (DNA).

unit is *N*-(2-aminoethyl)glycine. In contrast to DNA and RNA, PNA does not have any pentose sugar moieties or phosphate groups. For this reason, it is chemically stable and resistant to hydrolytic degradation and thus not expected to be broken down inside a living cell. PNA is therefore capable of sequence-specific recognition of DNA and RNA obeying the Watson–Crick hydrogen bonding rule, and the hybrid formed has very high thermal stability, which means that the melting point temperature of a PNA–DNA hybrid is higher than that of a DNA–DNA hybrid. Additionally, hybridization is

Structural formula 2.2: Ribonucleic acid (RNA).

independent of the ionic strength, salt concentration and pH of the hybridization solution.[5]

Structural formula 2.3: Peptide nucleic acids (PNA).

### 2.1.2.2 Locked Nucleic Acid

Locked nucleic acid (LNA) is another example of synthetic NA created by furanose ring modification.[6] LNA, also known as bridge nucleic acid, has a nucleotide containing a methylene bridge that connects the 2′-oxygen of ribose with the 4′-carbon (structural formula 2.4). LNA bases are linked by the same phosphate backbone found in DNA or RNA. LNA–RNA hybrids are among the strongest hybrids in terms of their modification of melting point temperature.[7]

Structural formula 2.4: Locked nucleic acids (LNA).

## 2.2 Applications of Nucleic Acid Based Biosensor Technologies

An overview of different NA-based biosensor technologies with some details is briefly summarized in Table 2.1. Selected applications of various sensor technologies[8–64] developed by using NAs—natural ones (DNA, RNA) and synthetic analogs (PNA, LNA)—are also discussed in the following sections, which are structured on the basis of the chemical characteristics of bioreceptors.

**Table 2.1** A summary of various sensor technologies recently developed by using nucleic acids; natural ones (DNA, RNA) and synthetic analogs (PNA, LNA).

| Nucleic acid | Method | | Sensitivity/working range | Ref. no |
|---|---|---|---|---|
| DNA | Electrochemical | SWV, Gold screen-printed electrodes | 10 nM | 8 |
| | | SWV, CV, DNA-doped conductive films | 10 pg or 1.6 fmol in 0.1 mL | 9 |
| | | CV, EIS, Pt working electrode | 23 nm film: 10.5 $\mu A\,cm^{-2}\,nM^{-1}$ | 10 |
| | | | 57 nm film: 3.0 $\mu A\,cm^{-2}\,nM^{-1}$ | |
| | | | 114 nm film: 1.7 $\mu A\,cm^{-2}\,nM^{-1}$ | |
| | | DPV, CV, Graphite screen-printed electrodes | 1 $\mu g\,mL^{-1}$ | 11 |
| | | DPV, SWV, PSA | — | 12 |
| | | LSV, CV, Gold electrode | 30 $pg\,mL^{-1}$ | 13 |
| | | EIS, CV, DPV, PbSe nanoparticle/chitosan composite films on CPE | $1.6 \times 10^{-11}$ M | 14 |
| | | CSV, Magnetic beads, HMDE, PyGE | | 15 |
| | | SWV, GC disks | 10 nM | 16 |
| | | SWV, EIS, GC disks | 0.03 nM | 17 |
| | | CV, EIS, Gold coated glass electrode | 115.8 $\mu A$/ng | 18 |
| | | SWV, GC disks and plates | — | 19 |
| | | CV, Pt chip electrode | Detection limit 1 nmole with sensitivity 0.62 $\mu A$/nmol | 20 |
| | | DPV, PGE | 74.8 $fmol\,mL^{-1}$ target concentration in the 50 $\mu L$ samples | 21 |
| | | DPV, CPE | 9.00 nM | 23 |
| | | CV, PGE | 0.25 $ng\,mL^{-1}$ or 53 pM | 24 |
| | | DPV, CV, CPE | $10^{-8}\,mol\,dm^{-3}$ | 25 |
| | | CV, Photolithography technique | | 26 |
| | | CV, SPCE | 500 $\mu g\,mL^{-1}$ | 27 |
| | | DPV, GC surface | dsDNA 50 $\mu g\,mL^{-1}$ | 28 |
| | | SWV, Carbon SPE | | 29 |
| | | Chronopotentiometry, CPE | 120 $ng\,mL^{-1}$ | 30 |

| | Technique/Electrode | Detection | Ref |
|---|---|---|---|
| | DPV, GCE | — | 31 |
| | DPV, PGE | Between 0.018 and 2.56 ppm | 32 |
| | EIS, GCE | 5 nM target DNA | 33 |
| | CV, NPV, EIS, Pt disk working electrode | 0.02 μM | 34 |
| | SWV, SPCE | 3.0 μg mL$^{-1}$ | 35 |
| | CV, Gold disk electrode | 0.1 mM | 36 |
| | DPV, ACV, AdTSV, HMDE, GCE, CPE | 1.85 ng mL$^{-1}$ ssDNA | 37 |
| | DPV, Platinum nanoparticles with MWCNTs | 2.35 ng mL$^{-1}$ DNA probe 2.03 ng mL$^{-1}$ hybrid 1.0×10$^{-11}$ mol L$^{-1}$ | 38 |
| | Chronopotentiometry, SPCE | 0.05 μM for 1,2-diaminoanthraquinone 0.2 μM for 2-anthramine 5 μM for 2-naphthylamine | 39 |
| | DPV, SPCE | 0.6 amol mL$^{-1}$ | 40 |
| | DPV, Gold electrodes | 0.72 μM | 41 |
| | SWV, SPCE | — | 42 |
| | DPV, MWNT modified GCE | 2.3×10$^{-4}$ mol L$^{-1}$ | 43 |
| | EIS, Gold electrode | 1.0 ng μL$^{-1}$ of the target DNA | 44 |
| | CV, nonfaradic EIS Gold electrode | 50 pmol of HBV DNA 160 pmol of HIV DNA in 20 L sample | 45 |
| | DPV, Nanogold modified electrode | 1.0×10$^{-11}$ M | 52 |
| | CV, SPCE | 2.0 μg mL$^{-1}$ ssDNA | 56 |
| | EIS | 1 nM | 22 |
| Mass sensitive | QCM, CdS nanoparticles, gold electrode | 1 nM | 22 |
| Optical | SPR, Gold sensor chip | range 10–20–50 ppm | 47 |
| | Enzymatic chemiluminescence detection Magnetic beads | 10$^{-17}$ mol DNA | 46 |
| | PL spectrum Crystalline silicon | 10 μM | 48 |
| | Fluorescence-based Silica optical fibers | — | 49 |
| | Spectrophotometric detection | 1.0 μL of target DNA T1 (1.0×10$^{-4}$ M) | 57 |

**Table 2.1** (continued)

| Nucleic acid | Method | | Sensitivity/working range | Ref. no |
|---|---|---|---|---|
| RNA | Electrochemical | CSV, HMDE, PGE magnetic beads | – | 15 |
| | | CV, Screen-printed carbon electrode modified with MWCNTs | $2.0\,\mu g\,mL^{-1}$ ssDNA | 56 |
| | | IDUAs | – | 60 |
| | Mass sensitive | QCM | – | 59 |
| | Optical | Spectrophotometric detection | $2.0\,\mu L$ target RNA ($1.0\times10^{-4}$ M) | 57 |
| | | SPR | – | 61 |
| | | Membrane-based DNA/RNA hybridization system using liposome amplification | 5 fmol | 58 |
| PNA | Electrochemical | SWV, GC disks | 10 nM | 16 |
| | | DPV, Nanogold modified electrode | $1.0\times10^{-11}$ M | 52 |
| | | EIS, Gold electrodes | The discrimination of the target DNA were investigated at three different concentrations: 50, 200, and 400 nM | 53 |
| | | DPV, Gold electrodes | $2.43\,ng\,mL^{-1}$ target | 51 |
| | | DPV | $10^{-6}$ M | 54 |
| | Optical | UV–VIS | $10^{-7}$ M | 54 |
| | | Reflectometric interference spectroscopy | $0.5\times10^{-3}\,\mu M$ | 55 |
| LNA | Electrochemical | CV, Nanogold modified GC electrode | $1.0\times10^{-11}$ M | 62 |
| | | DPV, Gold electrode | $1.2\times10^{-10}$ M | 63 |
| | | CV, SPEC | 78 pM | 64 |
| | | EIS | – | 50 |
| | Mass sensitive | QCM | – | 50 |

ACV, alternating current voltammetry techniques; AFM, atomic force microscopy; AdTSV, adsorptive transfer stripping voltammetry; AuE, gold electrode; cDNA, complementary DNA; CSV, cathodic stripping voltammetry; CPE, carbon paste electrode; DPV, differential pulsed voltammogram; dsDNA, double-stranded DNA; EIS, electrochemical impedance spectroscopy; FT-IR, Fourier transform infrared spectroscopy; GC, glassy carbon; HMDE, hanging mercury drop electrode; IDUAs, amperometric ultramicroelectrode arrays; LSPR, localized surface plasmon resonance; LSV, linear sweep voltammetry; MWCNTs, multi-walled carbon nanotubes; NPV, normal pulse voltammetry; ODN, oligonucleotide; PGE, pencil graphite electrode; PL, photoluminescence spectrum; PSA, potentiometric stripping analysis; PyGE, pyrolytic graphite electrode; QCM, quartz crystal microbalance; SPE, screen-printed electrode; SPCE, screen-printed carbon electrode; SPR, surface plasmon resonance; ssDNA, single-stranded DNA; SWV, square wave voltammetry.

## 2.2.1 DNA Biosensors

There have been many attempts to develop a number of DNA-based biosensors by using different transducers such as optical, electrochemical, mass sensitive, *etc.* in order to monitor DNA interactions and the specific DNA hybridization using short single-strand DNA sequences, *oligodeoxyribonucleotides (ODNs)*. In this section different examples of labeled and label-free electrochemical and optical approaches are reported.

Since the discovery of the electroactivity of NAs by Emil Palecek,[65] there have been many electrochemical approaches to analyzing NAs. As an example, Ricci *et al.*[8] used an hairpin ODN probe immobilized on a gold screen-printed electrodes as the basis for E-DNA sensors. An E-DNA sensor is based on hybridization-induced folding of an electrode-bound, redox-tagged DNA probe. The detection limit of this E-DNA sensor was found to be 10 nM in a working range of 10–200 nM using square wave voltammetry (SWV).

Moreover, direct electrochemical DNA sensor has been developed using the ultrathin films of a conducting polymer polypyrrole doped with an ODN probe for detection of a biowarfare pathogen, *variola major* virus, in 1.6 fmol of complementary oligonucleotide target in 0.1 mL in seconds by using chronoamperometry.[9]

In another study of the development of a polymer-based DNA sensor,[10] the authors prepared a simple and label-free electrochemical sensor for recognition of the DNA hybridization event using a functionalized conducting copolymer, poly[pyrrole-*co*-4-(3-pyrrolyl)butanoic acid].

Marazza *et al.*[11] introduced the concept of the *single-use electrochemical sensor* for the detection of DNA hybridization by using ODN immobilized onto graphite screen-printed electrodes. The hybridization event was monitored in their study by differential pulse voltammetry and chronopotentiometric stripping analysis using daunomycin hydrochloride as a hybridization indicator with a detection limit as 1 µg mL$^{-1}$ of target sequence.

The anticancer agent daunomycin was also used in the study of Sun *et al.*,[13] where a 24-mer ODN was covalently immobilized onto a self-assembled aminoethanethiol monolayer modified gold electrode. The authors used the daunomycin as an electroactive intercalator for electrochemical monitoring of the hybridization event, in combination with linear sweep voltammetry (LSV) with a detection limit of about 30 pg mL$^{-1}$.

In order to explore the steric effect on transduction during DNA hybridization, Piro *et al.*[19] monitored the electrochemical response versus the DNA target length for a constant DNA probe length. They reported that the current depends on the length of the double strand of DNA. The hybridization event was detected in their study by recording the modification of the redox process of the quinone group, using SWV.

DNA hybridization could also be performed onto the surface of commercially available magnetic particles.[15] A higher specificity of DNA hybridization could be achieved by the advantage of minimum nonspecific DNA adsorption. Palecek *et al.*[15] performed a magnetic assay with a high sensitivity and specificity while detecting relatively long target DNAs by using stripping voltammetry at mercury or solid mercury amalgam electrodes. The assay was based on the determination of purine bases, released from DNA by acid treatment. The same authors also used an enzyme-linked immunoassay of target DNA modified with osmium tetroxide 2,2′-bipyridine (Os,bipy) and carbon electrodes.

An electrochemical DNA sensor based on a sandwich hybridization assay[17] was developed for detection of target sequence related to lignin peroxidase (*lip*) genes using a gold electrode in combination with an amperometric method. In this study of Tang *et al.*, a monolayer of thiolated capture probe was prepared at the surface of gold electrode through self-assembly and the hybridization process was then carried out in the presence successively of target, biotinylated signalling probe and finally streptavidin–horseradish peroxidase (HRP) conjugate. The DNA conformation and surface coverage of the electrode were also characterized by impedance spectroscopy and squarewave voltammetry.

Manoj *et al.*[18] fabricated an electrochemical DNA hybridization sensor for the detection of meningitis in the presence of the electroactive indicator methylene blue (MB). The electrodes were first characterized using atomic force microscopy (AFM), Fourier transform infrared spectroscopy (FT-IR), electrochemical impedance spectroscopy (EIS) and cyclic voltammetric (CV) techniques, and then the hybridization event was detected in the presence of complementary DNA in the range of 7–42 ng $\mu L^{-1}$ in a hybridization time as short as 5 min.

Another application of an electrochemical DNA biosensor based on magnetic assay was developed for monitoring the DNA hybridization of wild-type hepatitis B virus (HBV) DNA in polymerase chain reaction (PCR) amplicons by Erdem *et al.*[21] Using a disposable graphite sensor, 20-mer synthetic oligonucleotides, and PCR amplicons of length 437 bp, the guanine oxidation signal observed at +1.0 V after DNA hybridization with a HBV probe was measured by differential pulse voltammetry (DPV), resulting in a detection limit of 74.8 fmole $mL^{-1}$ target concentration in 50 µL samples.

Tosar *et al.*[24] described two direct reagent-free detection methods using gold/polypyrrole/oligonucleotide modified electrodes by monitoring the decrease in the amplitude of polypyrrole oxidation and reduction peaks in cyclic voltammetry experiments.

Concerning the applications of electrochemical DNA biosensors for monitoring of chemical molecules interacting with DNA, Ferancová reported a study to detect pyrene derivatives.[25] After the combination of the cyclodextrin derivatives with polymer-modified screen-printed electrodes/heated carbon paste electrodes and DNA immobilization onto these electrodes, the voltammetric determination of 1-aminopyrene and 1-hydroxypyrene was finally performed at the electrode surface within the concentration range from $2 \times 10^{-8}$ to $4 \times 10^{-7}$ mol dm$^{-3}$ and from $2 \times 10^{-7}$ to $4 \times 10^{-6}$ mol dm$^{-3}$ with the limits of quantification down to $10^{-8}$ mol dm$^{-3}$.

There are a variety of optical DNA sensors measuring luminescence, fluorescence and refractive index. For example, the bacterial alkaline phosphatase *(phoA)* gene and HBV DNA were used as target DNA for the optical DNA sensor based on enzymatic chemiluminescence detection combined with magnetic particle assay in the study of Chen *et al*.[46] The detection cycle was less than 30 min, with detection limits for the *phoA* gene and HBV DNA at picogramme and femtomole level respectively.

In the study of Minunni *et al*.,[47] an optical DNA biosensor based on surface plasmon resonance (SPR) transduction was reported for drug screening. It was reported in this study that this sensor-based assay could be applied to the analysis of pure synthetic or natural molecules, and additionally to some fractions obtained by chromatographic separation of an extract of *Chelidonium majus* L. (greater celandine), a plant containing benzo[c]phenanthridinium alkaloids having intercalating properties. In the range of 10–50 ppm, the kinetic analysis of the interaction between the pure compounds and DNA was also studied by this SPR sensor.

An optical silicon-based label-free DNA sensor was developed by Di Francia *et al*.,[48] after n-type crystalline silicon wafers were electrochemically etched to form porous silicon layers and characterized in terms of porosity, pore distribution, surface composition, *etc*. It was reported in this study that the derivatized samples exhibited a photoluminescence that was stable in time and was not modified after exposure to the non-complementary DNA strand. Additionally, a measureable enhancement of the light emission was observed when the derivatized samples reacted with the complementary strand presenting that the specific single-stranded DNA/complementary DNA interaction could be recognized without using any further labeling steps.

According to the results based on fiber optic biosensors observed in the study of Watterson *et al*.,[49] the binding of fluorescein-labeled complementary DNA demonstrated that it was possible to achieve a selectivity coefficient of fully matched to single base pair mismatch of approximately 85 to 1, while maintaining less than 55% of the maximum possible signal that can be obtained from the fully matched target duplex.

## 2.2.2 RNA Biosensors

RNA probes for hybridization detection are less common. On the contrary, the specific monitoring of RNA by a sensor technology has a key importance especially when measuring pathogenic organisms. Thus, specific mRNA sequences can be represented a first-rate target molecule for the detection of viable pathogens.[15,56–61] Baeumner *et al*.[58] developed a highly sensitive and specific RNA biosensor for detection of the bacterium *Escherichia coli*, an important analyte for food and water safety. Viable *E. coli* was first identified and quantified by using a 200-nucleotide long target sequence from mRNA coding for a heat shock protein. Then, mRNA was extracted, purified and finally amplified using an isothermal amplification technique, NA sequence-based amplification. The biosensor quantified the amplified RNA by a membrane-based DNA/RNA hybridization system using liposome amplification and the detection limit was reported as 5 fmol per sample with the excellent correlation of as few as 40 *E. coli* cfu mL$^{-1}$.

A method for rapid sensitive detection of DNA or RNA was developed by using a composite screen-printed carbon electrode modified with multiwalled carbon nanotubes (MWNTs).[56] In this study, MWNTs showed catalytic characteristics for the direct electrochemical oxidation of G/A residues of single-stranded DNA (ssDNA) and A residues of RNA, with an accumulation time of 5 min. Ye *et al.*[56] reported that this biosensor could be used for detection of calf thymus ssDNA ranging from 17.0 to 345 µg mL$^{-1}$ with a detection limit of 2.0 µg mL$^{-1}$, and yeast tRNA ranging from 8.2 µg mL$^{-1}$ to 4.1 mg mL$^{-1}$.

Tedeschi *et al.*[59] reported a RNA biosensor developed by quartz crystal microbalance (QCM) technology. The binding ability of immobilized probes was tested evaluating their annealing behaviour with both complementary oligonucleotides and full-length target mRNA. The measurements of shift in QCM resonant frequency demonstrated that a label-free RNA biosensor can be developed with high specificity, reusability and the capability to provide quantitative information in a 50 nM solution of RNA target.

An electrochemical microfluidic biosensor with an integrated minipotentiostat for the quantification of RNA was developed by Kwakye *et al.*[60] based on target DNA or RNA hybridization, and liposome signal amplification. In this study, the reporter probe was coupled to liposomes entrapping the electrochemically active redox couple potassium ferri/ferrohexacyanide. After the capture probes were coupled to magnetic particles, the liposomes were then lysed to release the electrochemical markers that were detected electrochemically on an interdigitated ultramicroelectrode array in this biosensor just downstream of the magnet. This RNA biosensor was then demonstrated with the detection of dengue virus RNA.

As already mentioned, RNA oligonucleotides are not used for hybridization detection but they are frequently used for aptamer production. Aptamers are short ssDNA or RNA sequences that are selected *in vitro* based on their high affinity to a target molecule.[66] Oguro *et al.*[61] developed a RNA aptamer-immobilized sensor chip in combination with a surface plasmon resonance assay to detect the eukaryotic initiation factor 4A (eIF4A) at the nanogram level within whole-cell lysates after optimized sample preparation.

## 2.2.3 PNA Biosensors

PNA strands are electrically neutral, thus there is no electrostatic repulsion between two hybridized PNA or PNA–DNA strands. Accordingly, PNA–PNA as well as PNA–DNA duplexes have a higher association constant and better thermal stability than DNA–DNA duplexes.[3,4] In view of the fact that single-base mismatches can be selectively differentiated in a PNA–DNA duplex after developing the PNA-based hybridization sensors, there has recently been an increasing attention on the development of more selective PNA-based biosensors.[16,51-55]

PNA probes were attached covalently onto a quinone-based electroactive polymer.[16] An increase in the peak current of quinone was measured by SWV after a PNA–DNA hybridization process, with detection limit 10 nM. Additionally, this PNA biosensor was reported to be highly selective, especially for effectively detecting a single base mutation in the target sequence.

Ozkan et al.[51] showed that there was a selective hybridization between the covalently immobilized PNA probe and its target DNA, which was monitored by measuring the changes in the peak currents of MB as a hybridization indicator.

Fang et al.[52] described an electrochemical method for the detection of DNA–PNA hybridization using a ferrocene-functionalized polythiophene transducer, with a detection limit of $1.0 \times 10^{-11}$ M. In this study, PNA probes were immobilized onto the surface of a nanogold modified electrode resulting, with an increase in the amount of PNA probe immobilized and an increase of the electrical signal as well. It was also reported in the study of Fang et al. that there was a remarkable selective discrimination against noncomplementary DNA and four-base mismatched DNA by using PNA probes.

Cysteine-linked PNA assembled on gold electrodes was used for electrochemical sensing of PNA–DNA hybridization, and the electron transfer through the monolayers was investigated using EIS in the presence of target DNA and redox marker ions $[Fe(CN)_6]^{3-/4-}$ as hybridization indicator.[53]

Li et al.[54] reported the use of ferrocenyl azobenzene labeled PNA oligomers as effective electrochemical and photochemical probes.

## 2.2.4 LNA Biosensors

The development of LNA sensors based on different kinds of transducers[50,62-64] is still in progress. LNA-based biosensors offer very high affinity, making a perfect full hybridization with their complementary targets, and also extraordinary specificity to discriminate single-base mutations in the target sequence.[50]

Lin et al.[62] developed an electrochemical LNA biosensor for detection of the breakpoint cluster region gene and a cellular *abl* fusion gene in chronic myelogenous leukemia by using thiolated-hairpin locked NAs as the capture probe. Using DPV to measure the signal of the indicator (MB), the authors reported that the LNA probe, immobilized onto the surface of a nanogold/polyeriochrome black T film-modified glassy carbon electrode, could selectively hybridize with its target DNA.

In another study of Lin et al.,[63] a LNA probe was immobilized onto the surface of a gold electrode, and the hybridization between LNA probe and

DNA target was then monitored by measuring the changes at the peak current of MB in order to selectively detect the fusion gene in chronic myelogenous leukemia.

Magnetic particle assays based on enzymatic detection process were developed using DNA, PNA and LNA capture probes for electrochemical monitoring of specific hybridization by Laschi et al.[64]

## 2.3 Conclusion

Many types of biosensors have been investigated for analysis or quantification of NAs, and their interactions with other molecules. Thus, they play an important role for pharmaceutical, clinical, environmental and forensic applications. Various applications of different NA biosensors have been discussed in this chapter.

Recent progress in the development of novel biosensors, with the advances in nanotechnology, can provide useful methods to develop more sensitive and effective assays to monitor biorecognition process. Various applications of different NA biosensors have been overviewed in this chapter. Further improvements in their application, principally using chip technology, are urgently needed for environmental monitoring of low molecular weight pollutants, toxins, pathogens, *etc.*

## References

1. W. Saenger, *Principles of Nucleic Acid Structure*, Springer-Verlag, New York, 1984.
2. B. Lewin, *Genes VII*, Oxford University Press, New York, 2002.
3. D. P. Bartel, *Cell*, 2004, **116**, 281.
4. C. G. Liu, G. A. Calin, B. Meloon, N. Gamliel, C. Sevignani, M. Ferracin, C. D. Dumitru, M. Shimizu, S. Zupo, M. Dono, H. Alder, F. Bullrich, M. Negrini and C. M. Croce, *Proc. Natl. Acad. Sci. U. S. A.*, 2004, **101**, 9740.
5. M. Egolm, O. Buchardt, L. Christensen, C. Behrens, S. M. Freier, D. A. Driver, R. H. Berg, S. K. Kim, B. Norden and P. E. Nielsen, *Nature*, 1993, **365**, 566.
6. M. Petersen and J. Wengel, *Trends Biotechnol.*, 2003, **21**, 74.
7. K. Bondensgaard, M. Petersen, S. K. Singh, V. K. Rajwanshi, R. Kumar, J. Wengel and J. P. Jacobsen, *Chemistry*, 6, 2687.
8. F. Ricci, N. Zari, F. Caprio, S. Recine, A. Amine, D. Moscone, G. Palleschi and K. W. Plaxco, *Bioelectrochemistry*, 2009, **76**, 208.
9. E. Komarova, M. Aldissi and A. Bogomolova, *Biosens. Bioelectron.*, 2005, **21**, 182.
10. H. Peng, C. Soeller, N. Vigar, P. A. Kilmartin, M. B. Cannell, G. A. Bowmaker, R. P. Cooney and J. Travas-Sejdic, *Biosens. Bioelectron.*, 2005, **20**, 1821.

11. G. Marrazza, I. Chianella and M. Macsini, *Biosens. Bioelectron.*, 1999, **14**, 43.
12. F. Lucarelli, G. Marrazza, A. P. F. Turner and M. Macsini, *Biosens. Bioelectron.*, 2004, **19**, 515.
13. X. Sun, P. He, S. Liu, J. Ye and Y. Fang, *Talanta*, 1998, **47**, 487.
14. J. K. Xie, K. Jiao, H. Liu, Q. X. Wang, S. F. Liu and X. Fu, *Chin. J. Anal. Chem.*, 2008, **36**, 7.
15. E. Paleček, M. Fojta and F. Jelen, *Bioelectrochemistry*, 2002, **56**, 85.
16. S. Reisberg, L. A. Dang, Q. A. Nguyen, B. Piro, V. Noel, P. E. Nielsen, L. A. Le and M. C. Pham, *Talanta*, 2008, **76**, 206.
17. L. Tang, G. Zeng, G. Shen, Y. Li, C. Liu, Z. Li, J. Luo, C. Fan and C. Yang, *Biosens. Bioelectron.*, 2009, **24**, 1474.
18. M. K. Patel, P. R. Solanki, S. Seth, S. Gupta, S. Khare, A. Kumar and B. D. Malhotra, *Electrochem. Commun.*, 2009, **11**, 969.
19. B. Piro, S. Reisberg, V. Noel and M. C. Pham, *Biosens. Bioelectron.*, 2007, **22**, 3126.
20. J. Cha, J. I. Han, Y. Choi, D. S. Yoon, K. W. Oh and G. Lim, *Biosens. Bioelectron.*, 2003, **18**, 1241.
21. A. Erdem, D. O. Ariksoysal, H. Karadeniz, P. Kara, A. Sengonul, A. A. Sayiner and M. Ozsoz, *Electrochem. Commun.*, 2005, **7**, 815.
22. H. Peng, C. Soeller, M. B. Cannell, G. A. Bowmaker, R. P. Cooney and J. Travas-Sejdic, *Biosens. Bioelectron.*, 2006, **21**, 1727.
23. M. S. Hejazi, J. B. Raoof, R. Ojani, S. M. Golabi and E. H. As, *Bioelectrochemistry*, 2010, **78**, 141.
24. J. P. Tosar, K. Keel and J. Laíz, *Biosens. Bioelectron.*, 2009, **24**, 3036.
25. A. Ferancová, M. Bucková, E. Korgová, O. Korbut, P. Gründler, I. Wärnmark, R. Štepán, J. Barek, J. Zima and J. Labuda, *Bioelectrochemistry*, 2005, **67**, 191.
26. K. Hashimoto, K. Ito and Y. Ishimori, *Sens. Actuators B*, 1998, **46**, 220.
27. J. Hubalek, J. Prasek, D. Huska, M. Adamek, O. Jasek, V. Adam, L. Trnkova, A. Horna and R. Kizek, *Procedia Chem.*, 2009, **1**, 1011.
28. A. M. Oliveira-Brett and V. C. Diculescu, *Bioelectrochemistry*, 2004, **64**, 143.
29. F. Lucarelli, A. Kicela, I. Palchetti, G. Marrazza and M. Macsini, *Bioelectrochemistry*, 2002, **58**, 113.
30. J. Wang, G. Rivas, J. R. Fernandes, J. L. L. Paz, M. Jiang and R. Waymire, *Anal. Chim.Acta*, 1998, **375**, 197.
31. S. C. B. Oliveira, O. Corduneanu and A. M. Oliveira-Brett, *Bioelectrochemistry*, 2008, **72**, 53.
32. B. Dogan-Topal, B. Uslu and S. A. Ozkan, *Biosens. Bioelectron.*, 2009, **24**, 2358.
33. O. A. Arotiba, A. Ignaszak, R. Malgas, A. Al-Ahmed, P. G. L. Baker, S. F. Mapolie and E. I. Iwuoha, *Electrochim. Acta*, 2007, **53**, 1689.
34. Kh. Ghanbari, S. Z. Bathaie and M. F. Mousavi, *Biosens. Bioelectron.*, 2008, **23**, 1825.
35. F. Lucarelli, G. Marrazza, I. Palchetti, S. Cesaretti and M. Mascini, *Anal. Chim. Acta*, 2002, **469**, 93.

36. Y. Jin, X. Yao, Q. Liu and J. Li, *Biosens. Bioelectron.*, 2007, **22**, 1126.
37. D. Ozkan, P. Kara, K. Kerman, B. Meric, A. Erdem, F. Jelen, P. E. Nielsen and M. Ozsoz, *Bioelectrochemistry*, 2002, **58**, 119.
38. N. Zhu, Z. Chang, P. He and Y. Fang, *Anal. Chim. Acta*, 2005, **525**, 21.
39. G. Chiti, G. Marrazza and M. Mascini, *Anal. Chim. Acta*, 2001, **427**, 155.
40. F. Azek, C. Grossiord, M. Joannes, B. Limoges and P. Brossier, *Anal. Biochem.*, 2000, **284**, 107.
41. A. Mehdinia, S. H. Kazemi, S. Z. Bathaie, A. Alizadeh, M. Shamsipur and M. F. Mousavi, *J. Pharm. Biomed. Anal.*, 2009, **49**, 587.
42. F. Lucarelli, I. Palchetti, G. Marrazza and M. Macsini, *Talanta*, 2002, **56**, 949.
43. G. Cheng, J. Zhao, Y. Tu, P. He and Y. Fang, *Anal. Chim. Acta*, 2005, **533**, 11.
44. Y. Akagi, M. Makimura, Y. Yokoyama, M. Fukazawa, S. Fujiki, M. Kadosaki and K. Tanino, *Electrochim. Acta*, 2006, **51**, 6367.
45. W. M. Hassen, C. Chaix, A. Abdelghani, F. Bessueille, D. Leonard and N. Jaffrezic-Renault, *Sens. Actuators B*, 2008, **134**, 755.
46. X. Chen, X. E. Zhang, Y. Q. Chai, W. P. Hu, Z. P. Zhang, X. M. Zhang and A. E. G. Cass, *Biosens. Bioelectron.*, 1998, **13**, 451.
47. M. Minunni, S. Tombelli, M. Mascini, A. R. Bilia, M. C. Bergonzi and F. F. Vincieri, *Talanta*, 2005, **65**, 578.
48. G. D. Francia, V. L. Ferrara, S. Manzo and S. Chiavarini, *Biosens. Bioelectron.*, 2005, **21**, 661.
49. J. H. Watterson, P. A. E. Piunno and U. J. Krull, *Anal. Chim. Acta*, 2002, **457**, 29.
50. F. Lucarelli, S. Tombelli, M. Minunni, G. Marrazza and M. Mascini, *Anal. Chim. Acta*, 2008, **609**, 139.
51. D. Ozkan, A. Erdem, P. Kara, K. Kerman, J. J. Gooding, P. E. Nielsen and M. Ozsoz, *Electrochem. Commun.*, 2002, **4**, 796.
52. B. Fang, S. Jiao, M. Li, Y. Qu and X. Jiang, *Biosens. Bioelectron.*, 2008, **23**, 1175.
53. T. H. Degefa and J. Kwa, *J. Electroanal. Chem.*, 2008, **612**, 37.
54. J. Li, M. Chen, H. Zhang and S. Liu, *Inorg. Chem. Commun.*, 2008, **11**, 392.
55. K. Kroeger, A. Jung, S. Reder and G. Gauglitz, *Anal. Chim. Acta*, 2002, **469**, 37.
56. Y. Ye and H. Ju, *Biosens. Bioelectron.*, 2005, **21**, 735.
57. Y. Zhang, Z. Li, Y. Cheng and Y. Wang, *Talanta*, 2009, **79**, 27.
58. A. J. Baeumner, R. N. Cohen, V. Miksic and J. Min, *Biosens. Bioelectron.*, 2003, **18**, 405.
59. L. Tedeschi, L. Citti and C. Domenici, *Biosens. Bioelectron.*, 2005, **20**, 2376.
60. S. Kwakye, V. N. Goral and A. J. Baeumner, *Biosens. Bioelectron.*, 2006, **21**, 2217.
61. A. Oguro, T. Ohtsu and Y. Nakamura, *Anal. Biochem.*, 2009, **388**, 102.
62. L. Lin, J. Chen, Q. Lin, W. Chen, J. Chen, H. Yao, A. Liu, X. Lina and Y. Chen, *Talanta*, 2010, **80**, 2113.

63. L. Lin, X. Lin, J. Chen, W. Chen, M. He and Y. Chen, *Electrochem. Commun.*, 2009, **11**, 1650.
64. S. Laschi, I. Palchetti, G. Marrazza and M. Macsini, *Bioelectrochemistry*, 2009, **76**, 214.
65. E. Palecek, *Nature*, 1960, **188**, 656.
66. M. Mascini, in *Aptamers in Bioanalysis*, ed. M. Macsini, John Wiley & Sons, New York, 2009.

CHAPTER 3
# Genosensing Environmental Pollution

ILARIA PALCHETTI, GIOVANNA MARRAZZA AND
MARCO MASCINI

Dipartimento di Chimica, Università degli studi di Firenze,
50019 Sesto Fiorentino (Fi), Italy

## 3.1 Introduction

Numerous detection systems based on the hybridization between a nucleic acid target (DNA or RNA) and its complementary probe, which is present either in solution or on a solid support, have been described in literature. Many formats of homogeneous assays allowing the determination of DNA sequences in solution have been developed. However, they do not allow easily continuous monitoring and miniaturization. DNA biosensors and DNA microarrays offer promising alternatives to these methods since they allow continuous, possibly reusable, fast, sensitive and selective detection of DNA/RNA hybridization. Development of DNA hybridization biosensors (also termed genosensors) and DNA microarrays has increased tremendously over the past few years, as demonstrated by the number of scientific publications in this area and the commercial exploitation of some devices. *DNA microarrays* (also termed DNA chips or gene chips) are successful examples of DNA hybridization bioassay. DNA microarrays are made from glass, plastic or silicon supports and are made up of tens to thousands of 10–100 µm reaction zones onto which individual oligonucleotide sequences have been immobilized. They allow multiple

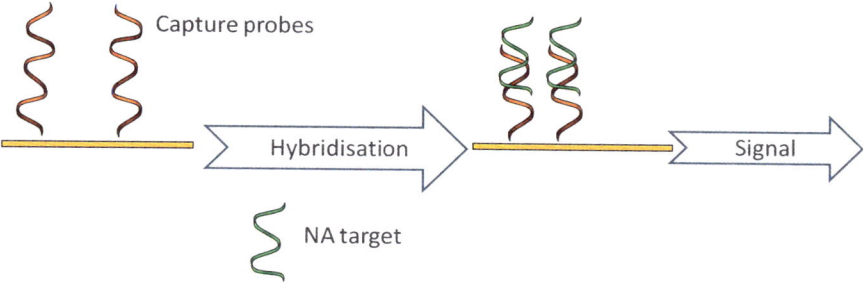

**Figure 3.1** Schematic representation of a genosensor.

parallel detection and analysis of the patterns of expression of thousands of genes in a single experiment. The transducer is generally a fluorescence microscope and is generally not in contact with the probe-modified surface. However, some examples of electrochemical microarrays are now on the market. Nevertheless, in this chapter we focus on genosensors, and the vast branch of literature on microarrays will be deliberately ignored.

DNA hybridization biosensors, also called genosensors, result from the integration of a sequence-specific probe (usually a short synthetic oligonucleotide, termed a *capture probe*) and a signal transducer (Figure 3.1). The capture probe, immobilized onto the transducer surface, acts as the biorecognition molecule and recognizes the target DNA or RNA.

This chapter addresses the issue of using genosensors for environmental monitoring. However, improvements in the application of genosensors in environmental pollution monitoring are to be expected in the near future as a result of the increasing number of different genosensor formats now being developed for other nonenvironmental targets. For these reasons, an overview of the important aspects that contribute to the creation of successful genosensing devices is given in the following sections. Probe design and immobilization, development and detection of the hybridization reaction will be discussed. Theoretically, all the transducing principles and techniques can be applied to the development of genosensors and several excellent reviews have been published on these topics in recent years.[1–13] These techniques will be presented here in a general form, not describing the theory, but only considering the features that are interesting for genosensing. Finally, examples of the application of genosensors in environmental monitoring will be described in a separate section.

## 3.2 Genosensor Development

Several different parameters have to be carefully controlled in order to develop a genosensor: probe design and immobilization, optimization and evaluation of the hybridization reaction, sample treatment. These critical aspects are discussed in the following sections.

### 3.2.1 Probe Design

The specificity of the hybridization reaction is essentially dependent on the biorecognition properties of the capture oligonucleotide; thus, design of the capture probe is undoubtedly the most important preanalytical step and a number of probes, varying in chemical composition and conformational arrangement, have been used to assemble a genosensor. Most commonly, the probes are linear oligonucleotides, either synthesized *in situ* or presynthesized and afterwards immobilized onto the sensor surface. However, structured (hairpin, pseudoknot) oligonucleotides are being used with increasing frequency.[13] Probes are commonly made up of DNA nucleotides; RNA is not often used because of its susceptibility to ribonuclease (RNase), a type of enzyme that catalyzes the degradation of RNA into smaller components. Recently, probes produced by chemical changes to the backbone of naturally occurring DNA or RNA, have been increasingly used in NA sensing techniques: locked nucleic acid (LNA) and peptide nucleic acid (PNA) are the most popular of these (see Chapter 2).

#### 3.2.1.1 Linear Probe

The design of linear probes now has the great advantage of decades of experience, which has led to much commercially available software. However, the design of full sets of arrayed probes for the screening identification of closely related and unrelated sequences, such as in the case of pathogens, still represents a particular challenge. For this purpose, once the genomic sequences have been retrieved from the relevant databanks and assembled and aligned using dedicated software, PCR primers and reporter oligonucleotides are typically selected within highly conserved regions of each bacterial genome. According to the requirements of each specific application, capture oligonucleotides can be then designed within either hypervariable or highly conserved regions. Candidate sequences are finally tested for theoretical melting temperature ($T_m$) and secondary structure formation, and searched for homologies using a basic local alignment search tool (BLAST). Use of capture probes in the order of 18–25 nucleotides long usually confers higher levels of specificity to the hybridization reaction. However, excessively long capture oligonucleotides often exhibit particularly unfavorable hybridization specificity. While a single or a few mismatches are unlikely to significantly destabilize even a 30-mer probe–target duplex over a wide range of experimental conditions,[14] the general hybridization efficiency of probes this length or longer might be very low, as a result of intramolecular hydrogen bonding and consequent formation of nonreactive structures.

#### 3.2.1.2 Structured Probe

Structured probes are oligonucleotides with stable secondary structures. *Hairpin* capture probes are mainly used in biosensor technology. They are oligonucleotides with a self-complementary sequence at both ends (stem) and a base-unpaired region (single-stranded loop) which contains the capture sequence. Another structure is a *pseudoknot*, which is a NA secondary structure containing at least two stem–loop structures in which half of one stem is intercalated between the two halves of another stem.

Proper design of the structured probes is obviously crucial, as functionality, selectivity and sensitivity of such capture oligonucleotides are reported to strongly depend on the amplitude of the loop and the length of the stem region.[15,16] In particular, some authors claim that hairpin probes have better selectivity than analogous linear oligonucleotides.[13] Such an improved discrimination capability has been attributed to the stem–loop structure of the probe, which stabilizes the dissociated state of the probe–analyte duplex especially in the presence of mismatched base pairs.[16] The temperature range within which discrimination between the two targets is possible is wider for structured probes than it is for the corresponding linear probes.[13] Structured probes are also attractive because of the possibility of self-tagging. As self-reporting probes, they allow direct and sensitive detection of unlabeled nucleic acid targets by reducing the number of steps in the set-up, and without the use of exogenous reagents.

However, other authors[15,17] showed different hairpin probes to undergo nonspecific hybridization with a number of sequences that presented a certain degree of homology with their target. These results raised the question of whether the hairpins are truly advantageous in terms of selectivity as immobilized capture probes.

## 3.2.2 Probe Immobilization

The ability to immobilize the probe in a predictable manner while maintaining its inherent affinity for the target NA is crucial to the overall performance of the device. Hence, independently of the molecular identity and conformation of the probe, some general aspects must be taken into account when considering its surface attachment. It is obvious that the choice of the most appropriate immobilization protocol is strictly dependent on the characteristics of the transducing material (*e.g.*, metal, glass, carbon surfaces). Robust immobilization chemistries are usually preferred, in order to prevent desorption of the probes from the sensing layer. Retention in a polymeric matrix, covalent attachment on a functionalized support, affinity immobilization (attachment of biotinylated probes to streptavidin-coated surfaces) and self-assembly (*e.g.*, chemisorption of thiol-modified probes onto gold surfaces) are, to date, the most successful approaches. This is mainly because each of these immobilization strategies can lead to ordered sets of properly oriented probes. Moreover, such chemistries allow control of the conformational freedom of the probes and the corresponding interchain space through modulation of the surface coverage. Hybridization efficiencies as high as 100% have even been reported in the most favorable cases, thus demonstrating that the resulting bioarchitectures are

**Table 3.1** Different procedures for probe immobilization on the transducer surface.

| Type | Principle | Advantage/Disadvantage |
|---|---|---|
| **Adsorption** | Direct physisorption onto transducer surface | Adsorption is the simplest immobilization method because it does not require any nucleic acid modification. However, besides being nonoriented, physisorbed molecules are limited in their mobility, which makes the hybridization event sterically inhibited. Efficient blockage of the surface (to prevent the nonspecific adsorption of DNA sequences, labels, *etc.*) can additionally be nontrivial. |
| **Covalent immobilization** | | |
| *Chemisorption* | The strong affinity of the thiol groups for noble metal surfaces enables the formation of covalent bonds between the sulfur and the gold atom: R-SH + Au → RSAu + $e^-$ + $H^+$ | The most recent approaches involved direct use of thiol tethered oligos for the formation of a SAM. As pioneered by the studies of Herne and Tarlov,[18] post-treatment of the probe-modified surface with a secondary thiol acting as "diluent" (*e.g.*, MCH) represented a simple means to control density and availability of the capture probe. The secondary thiol displaced the nonspecifically adsorbed probe molecules, while leaving the remaining ones in an upright position. An additional feature is the great stability of the surface-attached biolayer. |
| *Covalent attachment of a modified probe of functionalized surfaces* | Covalent reactions often use carbodiimide as a reagent, with or without NHS. EDC is the most frequently used activation coupling reagent. | Extremely versatile approaches that can be used for many different surfaces (carbon, glass, plastic, *etc.*) providing the presence of functional group onto the transducer surface. |

| | | |
|---|---|---|
| Avidin–biotin interactions | Another common procedure useful for silica surface is the silanization with a solution of APTES, the binding with glutaraldehyde (cross-linker) and the covalent attachment using a amine-terminated DNA-probe | |
| | Biotin is a small molecule that binds with a very high affinity to the avidin or streptavidin binding sites ($K_a = 10^{15}\,M^{-1}$). Moreover, avidin and streptavidin are tetrameric proteins that have four identical binding sites for biotin. Streptavidin with an isoelectric point (pI = 5) is thus preferred to avidin (pI = 10.5), to avoid non-specific interactions | The avidin (or streptavidin)–biotin complex is a very specific binding |
| Electropolymerization | The method involves the application of an appropriate potential to the working electrode soaked in an aqueous solution containing a biomolecule (e.g., DNA probe) and an electro-polymerizable monomer. Biomolecules present in the immediate vicinity of the electrode surface are physically incorporated in the growing polymer during its formation | Useful only for conductive surfaces |

APTES, 3-aminopropyltriethoxysilane; EDC, 1-ethyl-3-(3-dimethylaminopropyl)carbodiimide; pI isoelectric point; MCH, mercaptohexanol; NHS, N-hydroxysuccinimide; SAM, self-assembly monolayer.

fully accessible for interaction with the solution-phase target, while offering only minimal steric impediments. Nevertheless, adsorption has also been used for capture probe immobilization and different strategies based on this simple method are reported in literature (see Table 3.1).

However, functionalization of the sensing interface with the capture probe must follow two basic criteria. As a general rule, the probe has to be immobilized in such a way that its biorecognition capability is preserved as much as possible. Additionally, as a particular need of electrochemical detection schemes, the attached biolayer must not behave as a total insulator, thus allowing for electrochemical interrogation of the surface.

In the past few years many researchers have exploited the possibility of performing the hybridization event on the surface of paramagnetic micro- and nanobeads, to overcome the problem of absorption of nonspecific DNA sequences or labels at the transducer surface. This phenomenon can, in fact, reduce the sensitivity of the assay. Although such a concept of NA detection should perhaps be classified as a bioassay rather than a biosensor, the use of magnetic beads has gained popularity, notably for DNA hybridization sorting. The bead surface, coated either a thin gold layer, an amine-terminated organic shell, or streptavidin (or biotin), undergoes the formation of the capture probe monolayer through either gold thiolation or thioether, amide bridges or affinity linkage respectively. Following hybridization, the duplexes obtained were separated from the hybridization medium through magnetic extraction and washing, thus allowing enhanced specificity of the bioassay with the detection.

In addition, as described in Chapter 8, immobilization of capture probes onto nanomaterials has received increasing attention. Another area that raises the scientific interest in the development of immobilization strategies is the patterning of surfaces with micro- to nanometer resolution. More significantly for the fabrication of arrays, the accurate spatial localization of small amounts of molecules in well-defined areas, along with their specific properties, is of great importance for a multiplex analysis (Figure 3.2).

In this context, different research efforts have focused on increasing the number of different types of biological information per unit area. Localized electrochemistry appears to be a good candidate for direct addressing and patterning of conductive surfaces. In particular, scanning electrochemical microscopy (SECM) can be used for local electropolymerization of pyrrole and capture-probe-bearing pyrrole. SECM enables the electric field to be focalized, and was implemented for the design of arrays on raw conductive surfaces (Figure 3.2b). Localized electropolymerization could be also obtained mechanically through the use of mobile electrochemical microcells (Figure 3.2a).

## 3.2.3 Sample Treatment and Hybridization

The sensitivity of most hybridization-based biosensors does not generally allow the direct analysis of sequences. Detection thus depends on the preamplification

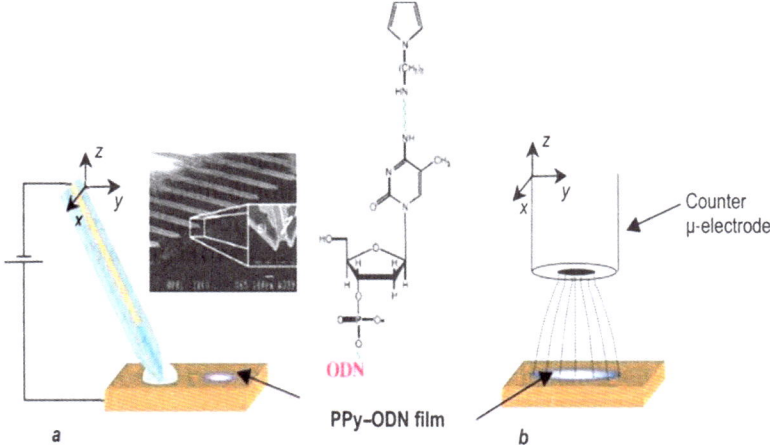

**Figure 3.2** (a) Cantilever-based electrochemical immobilization of oligonucleotide probes using electrocopolymerization of pyrrole and probe-bearing pyrrole moieties.[35] (b) Principle of electrochemical immobilization of probes using SECM in direct mode for the localized electrocopolymerization of pyrrole and probe-bearing pyrrole moieties.[40] Reproduced from ref. 4 by permission of the Royal Society of Chemistry.

of the target sequence and possibly on a further pretreatment of the amplified sample. Along with the use of capture probes with improved selectivity and hybridization efficiency, specific pretreatments of the samples can be used to greatly enhance the yield of the heterogeneous hybridization events and consequently the sensitivity of detection.

### 3.2.3.1 Sample Treatment

As already mentioned, detection depends on preamplification of the target sequence. This is true for most genoassays and especially for DNA detection, since RNA-based assays benefit from the typically high number of ribosomes per cell. The most commonly used preamplification techniques are the polymerase chain reaction or nucleic acid sequence-based amplification.

The *polymerase chain reaction (PCR)* is a technique widely used in molecular biology. It derives its name from one of its key components, a DNA polymerase used to amplify (*i.e.*, replicate) a piece of DNA by *in vitro* enzymatic replication. As PCR progresses, the DNA thus generated is itself used as template for replication. This sets in motion a chain reaction in which the DNA template is exponentially amplified. Generally, detection of particular DNA sequences, amplified by PCR, is carried out using gel electrophoresis. Although it is simple and effective for detection of PCR products under research conditions, gel electrophoresis is not considered to be a suitable method for routine analysis; moreover, gel electrophoresis cannot determine whether the sequence of the amplified target DNA is the same as that intended. In addition, ethidium bromide, which is a common staining agent used in gel electrophoresis, is carcinogenic. As alternative, genosensors can be used to control PCR amplification in a sequence-specific way. Amongst the different PCR variants, *multiplex PCR* is very useful as it allows the simultaneous detection of several organisms by introducing different primers to amplify DNA regions coding for specific genes of each bacterial strain targeted. *Real-time PCR* makes it possible to obtain quicker results without too much manipulation. This technique bases its detection in the fluorescent emission by a specific dye as it attaches itself to the targeted amplicon. One of the limitations of PCR techniques is that the user cannot discriminate between viable and nonviable cells because DNA is always present whether the cell is dead or alive. *Reverse transcriptase PCR (RT-PCR)* was developed in order to detect viable cells only.[19] This because the presence of mRNA can be regarded as a valid and convincing criterion for assessing cell viability. RT is an enzyme able to synthesize single-stranded DNA from RNA in the 5′–3′ direction. Several genes specifically present during the organism's growth phase can then be detected. Other PCR protocols are *asymmetric PCR*, which predominantly amplifies one DNA strand, and *direct PCR (DPCR)*, which allows amplification and detection of specific target nucleic acid sequences inside individual cells with no need for DNA extraction. Asymmetric PCR can result in higher sensitivity than symmetric PCR because of the presence in the PCR product of a high proportion of the single-stranded fragment, which can hybridize non-competitively with the probe. This also implies a faster hybridization of amplicons at the developed genosensor. Because of its speed, simplicity and minimal sample manipulation, DPCR has been demonstrated to be useful in detecting and quantifying bacteria in environmental samples, even from a sample containing one single cell. In this approach, the factor determining the amount of DNA available is cell lysis efficiency, which can be enhanced by employing methods to permeabilize the bacteria cell membranes to allow entry of reagents for amplification, and retard the diffusion of PCR products away from the cells, also not destroying their morphology or the microscale structure of the microbial community.

*Nucleic acid sequence-based amplification (NASBA)* is another genetic technique considered to be an interesting alternative to RT-PCR since it does not react with contaminating DNA. In recent years, quantitative NASBA assays have been developed for the detection of various viral and bacterial RNA in food samples.[20,21] NASBA technology is based on simultaneous enzymatic activity of reverse transcriptase, T7 RNA polymerase and RNase in combination with two oligonucleotides. It depends on selective primer-template recognition to drive a cyclical, exponential amplification of the target sequence.

Although high levels of fidelity generally characterize the PCR or NASBA amplification step, this process has always to be carefully optimized.

Nonspecific annealing of the primers, in PCR experiments, can, for example, lead to the generation of a wrong product (*i.e.*, a false-positive amplicon), while the presence of an inhibitor of polymerase can lead to a false-negative result. Given the possibility of directly sensing genomic DNAs, these drawbacks would all be overcome and the number of sample pretreatments reduced, thus significantly facilitating the detection of NAs. Nevertheless, scientists who have attempted the analysis of unamplified genomic DNA have often faced insurmountable difficulties. The huge size of genomic samples is the first of a series of obstacles. Hence, in order to facilitate the interfacial hybridization of such targets, fragmentation of the samples by sonication[22] or enzymatic digestion[23] has been suggested. Although simpler and cheaper, an undesired effect of sonication (which is a random process) might be disruption of the sequence targeted by the surface-immobilized probe. Nevertheless, even envisioning an optimal fragmentation, additional problems arise. If one considers that the genome of many organisms has a size of about $10^8$–$10^9$ bp and that fragments in the order of 500 bp on average would probably be the most suited for hybridization, as little as a single copy of the target would be "lost" between $10^5$ and $10^6$ nonrelevant strands. Diffusion of the fragment containing the target toward the probe-modified surface would be significantly slowed down by the concomitant presence of such sequences and further hindered by their electrostatic repulsion when accumulated in close proximity to the sensing layer. Effective match of the capture probe with the target sequence (the process that leads to duplex formation) is, therefore, a rather unlikely event. One has additionally to consider that, because of the complexity of genomic DNA, several of the fragments obtained could include sequences that present substantial homology with the true target. Yielding nonspecific hybridization with the capture probe, such sequences would therefore be responsible for false-positive signals. In other words analysis of genomic samples without amplification is still a big challenge for scientists.

Furthermore, in addition to amplification, many analytical protocols reported in literature require the samples to be pretreated prior to hybridization. The aims of such pretreatments include: (1) reduction of the complexity of the sample; (2) separation of possible interfering molecules (such as unincorporated primers and nucleotides); (3) generation of the target in a single-stranded form (or at least the minimization of amplicon sister strand reannealing); (4) disruption of thermodynamically stable secondary structures of the sequence (which might severely inhibit the interfacial biorecognition process). As a general rule, the number of sample pretreatments should be kept as small as possible, as they significantly slow down the whole analytical process.

### 3.2.3.2 Hybridization

Several variables affect the hybridization event at the transducer–solution interface and should be controlled carefully.[1] These experimental variables are

**Table 3.2** Advantages and disadvantages of different types of transducer used in DNA-based biosensors.

| Type | Principle | Advantage | Disadvantage |
|---|---|---|---|
| **Optical** | | | |
| Fluorescence | Detection of a fluorophore as label. Different format are possible including fluorescence microscope, fiber optics, FRET, molecular beacons, quantum dots, *etc*. | Sensitive, quantitative | Label-based techniques; equipment costly and not portable |
| SERS | Detection of a Raman label | A Raman dye can be either fluorescent or nonfluorescent, and a minor chemical modification of a dye molecule can lead to a new dye with a different Raman spectrum even if the two dyes exhibit virtually indistinguishable fluorescence spectra. Moreover, the spectral specificity of SERS is excellent in comparison to that of the fluorescence method | Label-based techniques; equipment costly and not portable. Moreover, SERS substrates must have an easily controlled protrusion size and reproducible structures |
| Chemiluminescence | Luminescent reactions catalyzed by enzyme or triggered by the application of a potential | Quantitative | Equipment costly and not portable |
| Colorimetry | Different formats are possible. Nowadays the use of gold nanoparticles may be used, as illustrated in Figure 3.3 | Almost any instrument can be used | In some cases only qualitative |
| Dual polarization interferometry | Nondiffractive optics to interrogate and resolve the size and density of DNA at a solid solution interface in real time | Label free, quantitative, real time measurements | Equipment costly and not portable |

# Genosensing Environmental Pollution

| | | | |
|---|---|---|---|
| SPR (an optical technique unaffected by the optical properties of the molecule) | SPR is used for determining refractive index changes at a surface. When light is incident on the prism side at a particular angle called the resonance angle, the intensity of the reflected light is at its minimum. In the presence of DNA on the metal (gold) surface, this angle variation is very sensitive. Changes in reflectivity give a signal that is proportional to the mass of the DNA bound to the surface. As the target binds to the probe, the mass and the refractive index increase. Different formats are possible like SPR imaging or SPFM | Label free, quantitative, real time measurements | Not enough sensitivity to monitor low molecular mass target such as short chain DNA molecules or very small packing density of the film cause very slight shift of the resonance angle. The sensitivity can be increased when using a label in SPR or SPFM format. |
| **Electrochemical** | | | |
| There are numerous labelled electrochemical DNA biosensors where the tag can be an enzyme, ferrocene, an interactive electroactive substance (a groove binder, or an intercalator), or nanoparticles. Other label-free electrochemical DNA biosensors also have been reported | Different detection technique can be used in both label and label-free formats. These include: Measurement of a current: amperometry, voltammetry (mainly SW or DPV), scanning electrochemical microscope Measurement of a current: amperometry, voltammetry (mainly SW or DPV), scanning electrochemical microscope Measurement of potential: Constant current potentiometry | High sensitivity, quantitative, small and portable transducers and instrumentation, low cost, and compatibility with micro-manufacturing technology | Solution is needed |

**Table 3.2** (*continued*)

| Type | Principle | Advantage | Disadvantage |
|---|---|---|---|
| **Piezoelectric** | A QCM sensor is a mass-sensitive sensor capable of measuring very small mass changes. It consists of a thin quartz disc sandwiched between a pair of electrodes. Quartz is a piezoelectric material that deforms when an electric field is applied across the electrode. The quartz crystal has a resonant frequency dependent on the total oscillating mass. This frequency increases with an increase in material on the QCM surface | Label free | Not enough sensitive to monitor low molecular mass molecules |
| SH-SAW | SH-SAW at the surface of a piezoelectric crystal, appears the most promising of all acoustic wave modes due to reduced damping effects associated with the presence of water | Label free | Sensitivity towards low molecular mass molecules |
| Microcantilever | These transducers are based on a response due to either surface stress variation or mass loading. Interaction between an immobilized ligand (e.g., a DNA probe) and an analyte (e.g., a DNA target) causes a change of the surface stress of the cantilever and can be detected as changes in the cantilever deflection | Label-free detection, high precision, reliability, reduced size, easy manufacture of multielement sensor arrays, and small thermal mass | Costly and not portable |

FRET, fluorescence resonance energy transfer; QCM, quartz crystal microbalance; SERS, surface-enhanced Raman scattering spectroscopy; SH-SAW, shear horizontal surface acoustic wave; SPR, surface plasmon resonance; SPFM, surface plasmon field-enhanced fluorescence spectroscopy.

referred to as *stringency*. Such variables typically include composition of hybridization and posthybridization washing buffers, for example salt concentration and pH; reaction temperature; presence of accelerating agents; and length of probe sequence. When dealing with more than a single probe at one time, the basic requirement for a functional system is the ability of all the different probes to hybridize their target sequences with high affinity and specificity under the same stringency conditions. This aspect makes the design of complex sets of probes even more difficult.

The design of probes for the analysis of samples susceptible to degradation (such as RNA) requires additional attention if a *sandwich hybridization scheme* is chosen. The selection of probes binding sites in close proximity to each other results is particularly convenient,[24] as it minimizes the adverse effects of sample degradation on the success of the sandwich hybridization reaction.

The hybridization reaction can be carried out in both a passive and an active mode. *Passive hybridization* takes place when temperature, buffer composition (salt concentration, formamide, *etc.*) and washes are used to control the stringency and rate of the reaction. In contrast to the passive approach, *active hybridization* uses electric fields on a microelectronic device to regulate nucleic acid transport, hybridization and stringency.[25] This technology, developed and implemented by Nanogen Inc. for arrays of 100 individually addressable microelectrodes, is reported to possess several advantages over passive methodologies. An electric field is first used to address different probes at specific sites of the microarray. The applied positive potential can also be used to increase the concentration of the target locally at the functionalized electrodes. Hence, while passive hybridization is limited by diffusion, electrochemical preconcentration of the target can ultimately lead to increased hybridization rates and efficiencies. Finally, reversal of the potential (*i.e.*, the application of negative voltages) imparts to the system an extremely high stringency for mutation detection based on the preferential destabilization of mismatched hybrids by means of purely electrostatic (repulsive) forces.

However, we would like to stress that it is always better to denature the sample prior to the hybridization step, either amplified by PCR or NASBA, or even digested by enzymes. This in order to have a single-stranded sample and to avoid the formation of secondary structure which can limit the hybridization efficiency. The denaturation of the target is commonly performed by thermal treatments, although chemical treatments are known.

### 3.2.4 Detection

Once the target NA has been captured onto the sensor surface, a range of different approaches can be used for transducing the biorecognition event (Table 3.2); these can be broadly divided into label-free and label-based schemes. *Label-free methods* translate the recognition behavior in a readable signal in real time by means of physicochemical changes in the transducer microenvironment subsequent to probe–target hybridization. By contrast, when organic and organometallic electroactive compounds, nanoparticles, catalytic and redox enzymes are permanently bound (*e.g.* covalently or via streptavidin–biotin interactions) to one of the constituents of the surface-tethered duplex, the method is considered to be *label-based*. Reagentless sensing concepts are the simplest, as nothing but the sample solution itself is needed to perform the analysis. Reagentless methods are not necessarily label-free. However, when both conditions are met (*i.e.*, the method is reagentless and label-free), one can take advantage of the fact that undesired effects, such as steric impediments to the hybridization reaction due to the reporter molecule, are completely absent. Such schemes have thus attracted intense research efforts because of their potential to provide analytical information more quickly, cheaply and safely than other strategies. As an additional advantage, transduction in the reagentless mode can also give access to the kinetics of the biorecognition event. Different methods of label-free transduction have been designed and rely on microgravimetric, electrical, electrochemical or optical transduction.

As already mentioned, when organic and organometallic electroactive compounds, nanoparticles, catalytic and redox enzymes are permanently bound (*e.g.* covalently or via streptavidin–biotin interactions) to one of the constituents of the surface-tethered duplex, the method is considered to be label-based. The sensitivity and reliability of label-based approaches are often still unrivaled, as the choice of these methodologies for the electrochemical microarray platforms now on the market also bears witness.

The label can be introduced into the target strand, also during the PCR amplification, or, to avoid modifying the target strand, a sandwich-type assay can be performed. In this latter case two hybridization steps take place. The target sequence is hybridized with both the capture probe and with a second probe (signaling probe), which is commonly labeled with an enzyme.

Finally, recent work has focused on the use of scanning probe microscopy, which enables the reading of highly densified DNA chips. Such a strategy, called here *DNA imaging*, opens the way to single-molecule interaction detection and represents a real challenge for the comprehensive study of biological interactions.

Obviously, owing to the intense research activity in this field, this review cannot provide a complete coverage of the tranduscton techniques reported in the DNA biosensor literature. In particular, fluorescence electrochemical electrical, optical (*e.g.* SPR), chemiluminescence, microgravimetric and microcalorimetric detection have been reported for DNA biosensors. However, this chapter merely tries to highlight recent interesting directions in terms of DNA immobilization or hybridization detection

### *3.2.4.1 Label-free Methods*

In label-free methods, the analytical signal relies on mass uptake, refractive index variation or space charge region modulation (*e.g.* dielectric constant, resistance).

*Microgravimetric transduction* of DNA hybridization is generally done using a quartz crystal microbalance (QCM). The QCM offers a mass detection at the nanogram level and thus allows accurate DNA determination, but without the possibility of multiparametric detection. As a consequence, an area of intense research activity is now focused on the development of new methodologies of microgravimetric transduction based on surface acoustic wave sensors or vibrating cantilevers. In particular, cantilevers can be used as resonators for the microgravimetric transduction of hybridization but they could be also involved as a nanomechanical detection tool when used in the deflection mode.

Detection based on *surface plasmon resonance (SPR)* is currently comparable to microgravimetric detection and is considered as an optical weighting tool. In such a transduction method, small modifications of refractive index in the vicinity of a gold-coated high-index glass substrate are sensed through a variation in the reflectivity of light reflected at the gold surface. A SPR instrument basically consists of a near-infrared diverging LED light source, a polarizer, a gold sensing layer, a reflecting mirror and a photodiode-array detector. The polarized monochromatic light is emitted toward the gold sensing surface and reflected at different angles. At certain angles of light incidence, resonance of the gold surface plasmons occurs and the intensity of the reflected light decreases dramatically. The light is reflected in a mirror and projected onto the photodiode array where the light intensity is measured. The position of the light intensity minimum is extremely sensitive to changes in refractive index (RI) of the fluid in the sensing area. Therefore, changes in RI near the sensing area can be measured by monitoring the shift of the light intensity minimum over time.[26]

*Electrochemical and electrical transduction* is the focus of much research effort, because of its low cost and power requirement, portability, independence of sample turbidity, ease of miniaturization and compatibility with manufacturing nanotechnologies. Electrical DNA sensors involve voltammetric and impedimetric detection as well as field-effect modulation based on field-effect transistors (FETs) where the change in surface charge subsequent to hybridization is detected. Amperometric and voltammetric methods are mainly based on either the intrinsic signal of DNA, such as the direct oxidation of adenine and guanine, or perturbations affecting the electrochemical behavior (conductivity, redox properties) of the capture probe matrix deposited onto the transducer. The former approach induces an irreversible process preventing multiple use, and is limited by the adenine and guanine content. Electroactive and conductive polymers are of great interest in this area, since they could be utilized, first of all, to capture probes to conductive surfaces through electropolymerization. In addition, unusual electronic or redox properties of polymer coatings could give them an active role in sensing hybridization.

In *impedimetric detection* based on local modifications of the space charge region due to the net immobilized charge increase consecutive to hybridization, the charge modification could be sensed either through differential capacitance measurements, through electrochemical impedance spectroscopy or by using FETs.

Again, as well documented in Chapter 8 of this book, the use of nanomaterials greatly improves the possibilities of label-free detection methods.

### 3.2.4.2 Label-based Methods

With a view to detecting extremely low concentrations of DNA, many publications have described signal amplification using the labeling of the duplex with redox probes, redox or chemiluminescent intercalators, passive nano-objects such as nanoparticles, or active markers such as enzymes or redox catalysts (Figure 3.3). Nanoparticles, including quantum dots and metal nanoparticles, have been widely used for the fabrication of DNA sensors since they can participate in the detection principle in a number of ways including fluorescence microscopy, conductive pathway generation, redox marker reservoirs or surface plasmon enhancement.

Instead of metal nanoparticles as redox label reservoirs, the use of enzyme markers constitutes an attractive strategy for time-controlled amplification via the catalytic production of redox species or light. Moreover, enzyme markers are conventionally used in bioassays and are thus easily available and well documented. For instance, many recent papers describe the use of alkaline phosphatase (ALP) and horseradish peroxidase (HRP) for the fabrication of electrochemical genosensors. These have an extremely sensitive detection limit due to their high turnover number (a molecule of HRP or ALP has a turnover frequency ranging from 1000 to 10 000 $s^{-1}$),[27] stability, low cost and broad substrate specificity.

Additionally, the use of a connecting matrix (conductive polymers) that provides direct electron transfer from the enzyme prosthetic centers to the electrode surface can be considered a helpful way to control the electron transfer from the enzyme active center to the electrode surface. Quite recently, as described in Chapter 5 of this book, DNAzymes have been proposed as labels in biosensing.

### 3.2.4.3 Imaging

These last years have seen increasing interest in scanning probe microscopy for DNA hybridization detection. Indeed, scanning probe techniques offer the advantage of high lateral resolution and thus allow a versatile approach of hybridization detection, from high-density biochip characterization to single biological event detection.

In this context, the *Kelvin probe force microscopy (KPFM)* was recently demonstrated as an interesting tool for measuring local variations in surface potential across a substrate of interest for high-resolution hybridization detection.[28] Since DNA strands are negatively charged, it was possible to measure the presence of a specific bound target on a DNA-modified surface without any form of labeling with a lateral resolution of 100 nm and a sensitivity of 50 nM. Moreover, KPFM was proved to furnish accurate measurements up to a scanning speed of 1.1 mm $s^{-1}$. This is quite interesting, since one important feature of scanning probe methodologies is with the reading speed, which could limit their application for the characterization of DNA hybridization on high-density biochips.

*Scanning electrochemical microscopy (SECM)* is also of great interest for biomolecular recognition detection. SECM methodology offers versatile detection principles (*e.g.*, positive or negative feedback modes together with collection mode), that are compatible with unlabeled hybridization detection as well as with redox amplification strategies of DNA hybridization. In this context, Toth *et al.*[29] reported preliminary studies for detecting DNA hybridization using glucose oxidase; the imaging of surface-confined DNA molecules, and hybridization through guanine oxidation induced by the tip generated Ru(bpy)$_3^{+}$ has been shown by Wang and Zhou[30] with probes immobilized on classical microarray glass slides. Zhou *et al.* proposed the use of SECM to image DNA hybridization with silver enhancement:[31] oligodeoxynucleotide (ODN) probes, immobilized on glass slides, were hybridized with biotinylated target, and hybridization was developed by a silver staining process (adsorption of streptavidin–gold nanoparticles followed by silver particle deposition) with consequent increase of surface conductivity. Komatsu *et al.* used SECM to visualize DNA duplex regions on a DNA microarray glass slide, through the use of ferrocenylnaphthalene diimide as an electrochemically active DNA hybrid indicator.[32] Schuhmann and coworkers reported a labelfree electrochemical detection scheme:[33,34] synthetic ODN were spotted on conducting surfaces and investigated by SECM in electrolytes containing [Fe(CN)$_6$]$^{3-/4-}$ as a negatively charged redox mediator. Under these conditions, significant decreases in positive feedback currents were observed above DNA-modified regions due to local appearance of repulsion between deprotonated phosphate groups of the immobilized DNA strands and the [Fe(CN)$_6$]$^{3-/4-}$ anions. Recently Fortin *et al.*[35] as well as Palchetti *et al.*[36] described SECM approaches based on the local alteration of conductivity due to an enzymatic product, using HRP and ALP respectively as a label for the surface-tethered duplex.

However, the resolution of SECM is directly dependent on the electrode size and may be pushed toward the detection of single biomolecular events. In such a context, Wang *et al.*,[37] by using electrochemical atomic force microscopy, explored the motional dynamics of end-grafted oligonucleotides and demonstrated DNA hybridization detection at the scale of 200 grafted molecules.

Moreover, in recent years SPR imaging has also been proposed as an interesting sensitive and label-free approach to DNA imaging. The SPR imaging technology is an SPR method of visualizing the whole of the biochip via a video CCD camera. This design enables the biochips to be prepared in an array format with each active site (spot) providing SPR information simultaneously.[38,39]

## 3.3 Genosensors for Environmental Monitoring

As already mentioned, interest in DNA-based biosensors has been growing in recent years. The development of systems allowing DNA detection is motivated

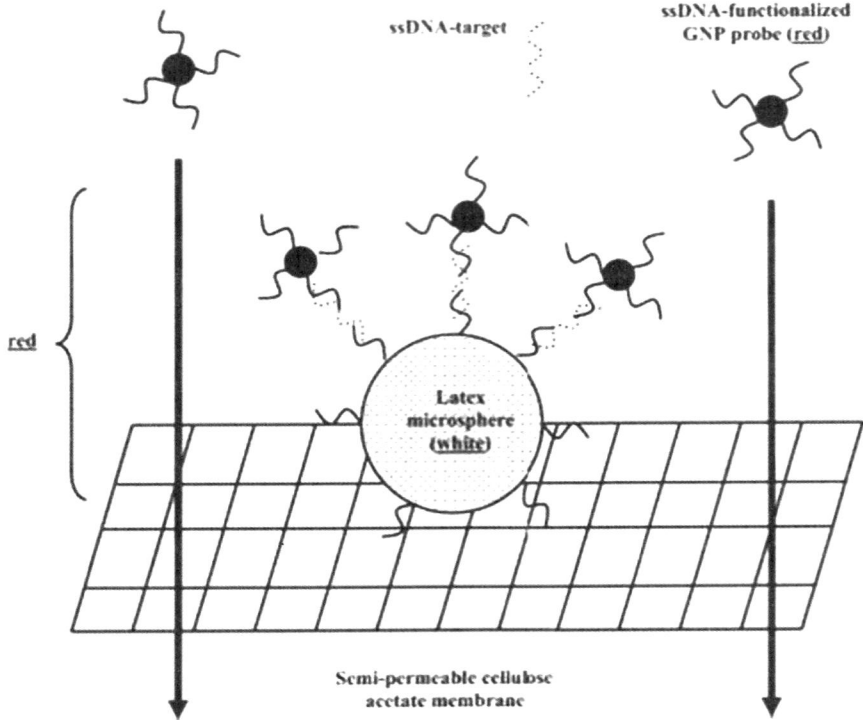

**Figure 3.3** Layout for DNA colorimetric detection with GNPs and latex microspheres. Free ssDNA-functionalized GNPs move freely across the semipermeable cellulose acetate membrane. Latex microparticles (white), on the other hand, are too large to pass through that barrier. In the presence of an ssDNA target, GNPs bind latex particles, generating large conjugates, which are retained by the membrane.[66] GNPs, gold nanoparticles.

by applications in many fields: DNA diagnostics, gene analysis, fast detection of biological warfare agents, and forensic applications. Detection of genetic mutations at the molecular level opens up the possibility of performing reliable diagnostics even before any symptom of a disease appears. Environmental applications such as pollution are also benefiting from genosensing technology.

The term *environmental pollution* covers many different forms of pollution, from chemical and physical pollution to microbiological pollution, including problems related to biodiversity, such as the loss of some plant and animal species or the introduction of new species by anthropogenic activity. The main environmental applications of genosensors can be obviously allocated to the field of species identification. For instance, genosensors have been extensively exploited in the detection of pathogenic microorganisms relevant to food, biodefense and environmental contamination. Mainly, genosensors have been coupled to PCR or NASBA as specific detection methods for the amplified base sequence. Amplification of the DNA target is necessary due to the low abundance and extreme complexity of the nonamplified targets. Recently, a QCM DNA biosensor based on nanoparticle amplification for the detection of the *E. coli* O157:H7 *eaeA* gene was proposed. The method, coupled to asymmetric PCR with biotin-labelled primers, was able to detect $2.67 \times 10^2$ cfu mL$^{-1}$.[41] In this case, a thiolated ssDNA probe specific to the *E. coli* O157:H7 *eaeA* gene was immobilized on the QCM surface. After bovine serum albumin blocking, the probe was exposed to the biotinylated target DNA; to amplify the frequency change, streptavidin-conjugated $Fe_3O_4$ nanoparticles were attached to the target. Mo *et al.*[42] reported the detection of *E. coli* cells in water combining the detection of the *lac* gene with PCR amplification and QCM detection. The QCM biosensor was based on a *lacZ*-specific thiolated ssDNA probe able to detect 1 mg mL$^{-1}$ target DNA and several viable *E. coli* cells in 100 mL of water. QCM detection of PCR samples of *Aeromonas* strains, isolated from water, plant or human specimens, has also been reported.[43]

Miranda-Castro *et al.* exploited the features of a surface-immobilized hairpin probe for the selective identification of *Legionella pneumophila* sequences.[16] A thiolated hairpin, immobilized onto a gold electrode surface, was combined with a sandwich-type hybridization assay in the presence of a biotinylated signaling probe and streptavidin–alkaline phosphatase as the reporter enzyme. The sensor allowed selective discrimination between *L. pneumophila* with a detection limit of 340 pM of *L. pneumophila* DNA. Direct comparison of the analytical performance of this assay with that of an analogous test based on the use of linear oligonucleotide probes demonstrated the superior sensitivity and selectivity of the hairpin oligonucleotide. In a further work this group tested the assay using disposable electrodes.[44]

The simultaneous electrochemical detection of *Salmonella enterica*, *Lysteria monocytogenes*, *Staphylococcus aureus*, and *E. coli* O157:H7 amplicons in less than 1 hour was demonstrated using a disposable gold electrode array.[45] Wang[46] described electrochemical genosensors for *Cryptosporidium* as well as *E. coli*, *Giardia* and *Microbacterium tuberculosis*.

Lermo *et al.*[47] developed a genomagnetic assay for the detection of food pathogens based on a graphite–epoxy composite magneto electrode (m-GEC) as electrochemical transducer. The assay was performed in a sandwich format by double-labeling the amplicon ends during PCR with a biotinylated capture probe, to achieve the immobilization on streptavidin-coated magnetic beads and a digoxigenin signaling probe, to achieve further labeling with the enzyme marker (anti-digoxigenin horseradish peroxidase). An update of that work was proposed,[48] where PCR specific amplification of the *E. coli* O157:H7 *eaeA* gene was detected using doubly-labeled amplicon instead of the use of the signaling probe. During PCR the target sequence was, in fact, amplified and labeled with biotin and digoxigenin, in order to achieve the immobilization and the enzymatic labeling respectively. The bacteria amplicons were then detected by using two different electrochemical genosensing strategies, a highly selective biosensor based on a bulk-modified avidin biocomposite (Av-GEB) sensor and the m-GEC. Using the proposed strategy, after 10 PCR cycles 4.5 and 0.45 ng mL$^{-1}$ of the original bacteria genome were detected with Av-GEB and m-GEC respectively.

The same group developed another procedure based on the use of the m-GEC described above.[49] The assay combines immunomagnetic separation (IMS), double-tagging PCR, and electrochemical magneto-genosensing of the double-tagged amplicon for *Salmonella enterica* serovar *Typhimurium*. The *Salmonella* was captured on magnetic beads modified with an antibody specific for *Salmonella* spp. The confirmation of the attached bacteria on the magnetic beads was performed by amplifying the genomic DNA with a double-tagging set of primers specific for *Salmonella*. During the PCR, not only the amplification of pathogenic bacteria genome was achieved, but also the double tagging of the amplicon ends with (1) the biotinylated capture primer to achieve the immobilization on streptavidin-modified magnetic beads and (2) the digoxigenin signaling primer to achieve the enzymatic detection through AntiDig-HRP reporter. The double-tagged amplicon was sensitively detected by and electrochemical magneto-genosensor. The enzymatic labeling was done by using antiDig-HRP. The authors report that this strategy was able to detect as low as 1 cfu mL$^{-1}$ of bacteria in Luria–Bertani buffer (LB) as well as well as in milk diluted 1/10 in LB, without showing any matrix effect. Moreover, the method was not affected by the food matrix, because of the use of magnetic beads. The authors conclude that the magnetic immunoseparation was able to effectively replace the selective plating, while the double-tagging PCR strategy with electrochemical genosensing, the biochemical probes and the serological confirmation reduced the time of the assay from 3–5 days to 3.5 h. They also pointed out that the IMS/double-tagging PCR/m-GEC electrochemical genosensing approach was able to give positive signal with dead or injured whole cells, but DNA released during food processing was not detected, unlike commercial PCR approaches without IMS. Moreover, PRC inhibitors were also avoided. If the sample was pre-enriched for 6 h in LB, as low as 0.04 cfu mL$^{-1}$ of *Salmonella* can be feasibly detected. Obviously, the same approach could also be designed to detect different *Salmonella* or *E. coli* serotypes by selecting a specific pair of primers or antibody.

Magnetic beads were also used as solid support for capture probe immobilization by Loaiza *et al.*[50] In this paper, direct asymmetric PCR was used to improve the sensitivity of the specific detection of a gene related to the *Enterobacteriaceae* bacterial family. Direct asymmetric PCR was combined with disposable magnetic DNA sensors based on the coupling of streptavidin–peroxidase to biotinylated *lacZ* gene target sequences. Tetrathiafulvalene (TTF)-modified gold screen-printed electrodes (AuSPEs) were used as transducers for the amperometric measurement (Figure 3.4).

These examples demonstrate interesting features; however, a significant challenge for all biosensor systems is minimizing sample preparation requirements, operational complexity and time. Microfluidic-based platforms show great potential in responding to this demand.

Electrochemical detection of harmful algae and other microbial contaminants in coastal waters using commercially available hand-held biosensors has also been demonstrated.[51,52] Electrochemical assays were developed for a red tide dinoflagellate (*Karenia brevis*), bacteria indicating fecal contamination

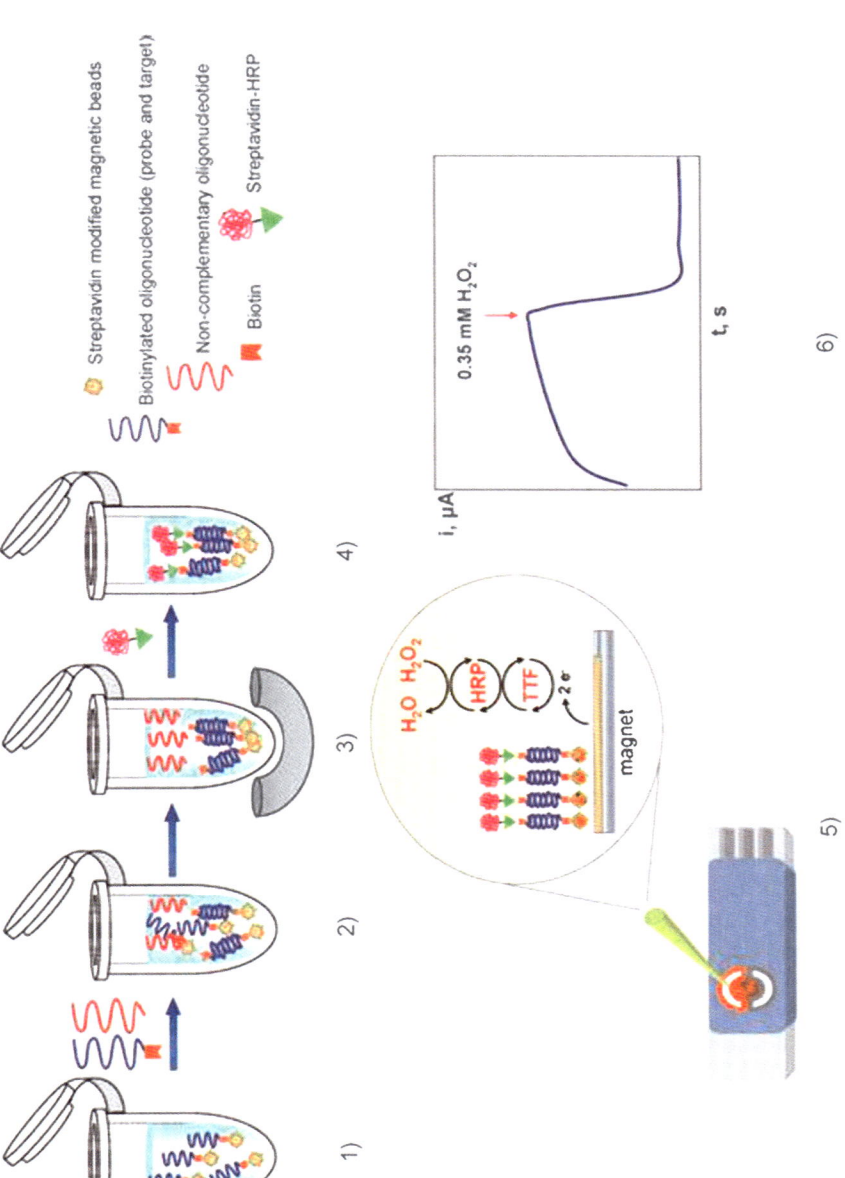

**Figure 3.4** Schematic representation of a procedure based on magnetic beads: (1) washing of probe-modified magnetic beads; (2) hybridization with the target *lacZ* gene probe; (3) separation of hybrid-modified magnetic beads and extraction of noncomplementary oligonucleotide; (4) enzymatic labeling with streptavidin–HRP; (5) deposition of hybrid-modified magnetic beads on the TTF-Au/SPEs; (6) amperometric detection of the mediated reduction of $H_2O_2$ with TTF. Reproduced from ref. 50 by permission of the Royal Society of Chemistry.

(*Enterococcus* spp.), markers indicative of human sources of fecal pollution (human cluster Bacteroides and the *esp* gene of *Enterococcus faecium*), bacterial pathogens (*Escherichia coli* O157:H7, *Salmonella* spp., *Campylobacter jejuni, Staphylococcus aureus*), and a viral pathogen (adenovirus).[52] For *K. brevis*, two assay formats (Rapid PCR-Detect and Hybrid PCR-Detect) were tested and both provided detection limits of 10 genome equivalents for DNA isolated from *K. brevis* culture and amplified by PCR. Sensitivity with coastal water samples was sufficient to detect *K. brevis* that was "present" (61 000 cells/L) without yielding false positive results and the electrochemical signal was significantly different than for samples containing cells at "medium" concentrations (100 000 to $<10^6$ cells/L). Detection of *K. brevis* RNA was also shown. Multi-target capability was demonstrated with an 8-plex assay for bacterial and viral targets using isolated DNA, natural beach water spiked with human feces, and water and sediments collected from New Orleans, Louisiana following Hurricane Katrina (2005). Furthermore, direct detection of dinoflagellate and bacterial DNA was achieved using lysed cells rather than extracted nucleic acids, allowing streamlining of the process. It was proposed that the methods could be used to rapidly (3–5 h) screen environmental water samples for the presence of microbial contaminants, and have the potential to be integrated into semi-automated detection platforms.

Metfies *et al.* reported the detection of the toxic alga *Alexandrium ostenfeldii*, responsible for shellfish poisoning, using a digoxygenin/anti-digoxygenin–HRP conjugate labeling scheme.[53] Algal rRNA was detected by sandwich hybridization in the presence of a digoxygenin-labeled reporter oligonucleotide. Labeling with HRP was achieved using a HRP-conjugated anti-digoxigenin antibody. The enzyme was oxidized in catalyzing the reduction of hydrogen peroxide to water, but the reduced form of HRP was readily regenerated by *p*-aminodiphenylamin, which acted as a mediator. Reduction of the oxidized mediator at –150 mV *versus* Ag/AgCl provided the electrochemical readout.

Baeumner *et al.*[54] detected as few as 40 *E. coli* cells per 1-mL sample using a simple optical dipstick-type biosensor coupled to NASBA, emphasizing the fact that only viable cells are detected and no false-positive signals are obtained from dead cells present in a sample, which is important with respect to safety assessments and also for food and environmental sample sterilization. Similarly, a biosensor for the protozoan parasite *Cryptosporidium parvum* was developed.[55] Baeumner's group also developed a genosensor with integrated microfluidic system and a minipotentiostat for the quantification of dengue virus RNA.[56] By combining microelectronics and microfluidics with the simple and effective liposome signal enhancement technology and an amperometric transducer, the authors designed a miniaturized electrochemical detection system (miniEC) that was easy to assemble and use. DNA and RNA molecules were quantified using a sandwich hybridization assay similar to the one previously described,[57,58] with the transduction mechanism being based on electrochemical rather than fluorescence detection.

A rapid procedure that did not require any amplification of the targeted nucleic acids by PCR or NASBA was reported by Elsholz et al.[59] Low-density electrical 16S ribosomal RNA (rRNA) specific oligonucleotide microarrays and an automated analysis system for the identification and quantization of pathogens such as *Escherichia coli*, *Pseudomonas aeruginosa*, *Enterococcus faecalis*, *Staphylococcus aureus*, and *Staphylococcus epidermidis* were described. The detection of bacterial cell-extracted 16S rRNA using SPR imaging of DNA arrays was also demonstrated.[60]

Scanning electrochemical microscopy was applied to the detection of a *Salmonella* PCR amplicon in Palchetti et al.[36]

Genosensors were also applied to the detection of genetically modified plants or foods, such as the SPR genosensors described by Feriotto et al. for the detection of Round-up ready soybeans in which PCR-amplified sequences from transgenic and wild-type soybeans were investigated by the SPR sensor principle.[61] Examples of QCM or electrochemical biosensors are reported in refs 62,63.

An edible fish species, *Lophius budegassa* (monkfish), was discriminated from the toxic *Takifugu niphobles* (pufferfish) by an electrochemical genosensor analysing the polymorphisms of the gene that encodes the 16S ribosomal subunit of their mitochondrial DNA.[64]

In a recent study, species specificity was detected in pork, chicken and bovine meats using loop-mediated isothermal amplicon (LAMP) and disposable electrochemical printed (DEP) chips. LAMP is a nucleic acid amplification method that amplifies target DNA with high specificity, efficiency and rapidity under isothermal condition (63 °C). An electrochemical genosensor with the DEP chips detects the amplicons by linear sweep voltammetry (LSV) observation of DNA–Hoechst 33258 interaction on the chip surface.[65]

## 3.4 Conclusions

The examples of environmental applications of genosensors reported here demonstrate that the interest of the use of these devices in this field is notable. A further increase of environmental applications is expected in the near future, resulting from the improvements in nanotechnology and nanoengineering areas that will lead to the development of integrated PCR devices. In this way gene analysis will become more "user friendly" and suitable for use outside the laboratory.

## References

1. J. Watterson, P. A. E. Piunno and U. J. Krull, *Anal. Chim. Acta*, 2002, **469**, 115.
2. R. Levicky and A. Horgan, *Trends Biotechnol.*, 2005, **23**, 143.
3. F. Lucarelli, S. Tombelli, M. Minunni, G. Marrazza, A. P. F. Turner and M. Mascini, *Anal. Chim. Acta*, 2008, **609**, 139.

4. S. Cosnier and P. Mailley, *Analyst*, 2008, **133**, 984.
5. A. Sassolas, B. D. Leca-Bouvier and L. J. Blum, *Chem. Rev.*, 2008, **108**, 109.
6. E. Palecek and M. Fojta, *Anal. Chem.*, 2001, **73**, 74A–83A.
7. F. R. R. Teles and L. P. Fonseca, *Talanta*, 2008, **77**, 606.
8. T. G. Drummond, M. G. Hill and J. K. Barton, *Nat. Biotechnol.*, 2003, **21**, 1192.
9. R. Moeller and W. Fritzsche, *IEEE Proc. Nanobiotechnol.*, 2005, **15**, 247.
10. J. Wang, *Biosens. Bioelectron.*, 2006, **21**, 1887.
11. M. Mascini, I. Palchetti and G. Marrazza, *Fresenius J. Anal. Chem.*, 2001, **369**, 15.
12. I. Palchetti and M. Mascini, *Analyst*, 2008, **133**, 846.
13. R. Miranda Castro, N. de los Santos Alvarez, M. J. Lobo Castanon, A. J. Miranda Ordieres and P. Tunon Blanco, *Electroanalysis*, 2009, **2077**.
14. P. Abad-Valle, M. T. Fernandez-Abedul and A. Costa-Garcia, *Biosens. Bioelectron.*, 2007, **22**, 1642.
15. D. M. Jenkins, B. Chami, M. Kreuzer, G. Presting, A. M. Alvarez and B. Y. Liaw, *Anal. Chem.*, 2006, **78**, 2314.
16. R. Miranda-Castro, P. de-los-Santos-Álvarez, M. J. Lobo-Castano, A. J. Miranda-Ordieres and P. Tunon-Blanco, *Anal. Chem.*, 2007, **79**, 4050.
17. F. Ricci, R. Y. Lai, A. J. Heeger, K. W. Plaxco and J. J. Sumner, *Langmuir*, 2007, **23**, 6827.
18. T. M. Herne and M. J. Tarlov, *J. Am. Chem. Soc.*, 1997, **119**, 8916.
19. S. Yaron, *J. Appl. Microbiol.*, 2002, **92**, 633.
20. S. R. Nugen, B. Leonard and A. J. Baeumner, *Biosens. Bioelectron.*, 2007, **22**, 2442.
21. N. Cook, *J. Microbiol. Methods*, 2003, **1761**, 1.
22. T. L. Mann and U. J. Krull, *Biosens. Bioelectron.*, 2004, **20**, 945.
23. F. Patolsky, A. Lichtenstein and I. Willner, *Nat. Biotechnol.*, 2001, **19**, 253.
24. K. Metfies, S. Huljic, M. Lange and L. K. Medlin, *Biosens. Bioelectron.*, 2005, **20**, 1349.
25. E. A. Barlaan, M. Sugimori, S. Furukawa and K. Takeuchi, *J. Biotechnol.*, 2005, **115**, 11.
26. J. H. T. Luong, K. B. Male and J. D. Glennon, *Biotechnology Advances*, 2008, **26**, 492.
27. F. Mallard, G. Marchand, F. Ginot and R. Campagnolo, *Biosens. Bioelectron.*, 2005, **20**, 1813.
28. A. K. Sinensky and A. M. Belcher, *Nat. Nanotechnol.*, 2007, **2**, 553.
29. R. E. Gyurcsanyi, G. Jagerszki, G. Kiss and K. Toth, *Bioelectrochemistry*, 2004, **63**, 207.
30. J. Wang and F. Zhou, *J. Electroanal. Chem.*, 2002, **537**, 95.
31. J. Wang, F. Song and F. Zhou, *Langmuir*, 2002, **18**, 6653.
32. F. Turcu, A. Schulte, G. Hartwich and W. Schuhmann, *Biosens. Bioelectron.*, 2004, **20**, 925.
33. F. Turcu, A. Schulte, G. Hartwich and W. Schuhmann, *Angew. Chem. Int. Ed.*, 2004, **43**, 3482.

34. M. Komatsu, K. Yamashita, K. Uchida, H. Kondo and S. Takenaka, *Electrochim. Acta*, 2006, **51**, 2023.
35. E. Fortin, P. Mailley, L. Lacroix and S. Szunerits, *Analyst*, 2006, **131**, 186.
36. I. Palchetti, S. Laschi, G. Marrazza and M. Mascini, *Anal. Chem.*, 2007, **79**, 7206.
37. K. Wang, C. Goyer, A. Anne and C. Demaille, *J. Phys. Chem. B*, 2007, **111**, 6051.
38. S. Szunerits, N. Knorr, R. Calemczuk and T. Livache, *Langmuir*, 2004, **20**, 9236.
39. C. Corne, J. B. Fiche, D. Gasparutto, V. Cunin, E. Suraniti, A. Buhot, J. Fuchs, R. Calemczuk, T. Livache and A. Favier, *Analyst*, 2008, **133**, 1036.
40. E. Descamps, T. Leichle, B. Corso, S. Laurent, P. Mailley, L. Nicu, T. Livache and C. Bergaud, *Adv. Mater.*, 2007, **19**, 1816.
41. X. Mao, L. Yang, X. L. Su and Y. Li, *Biosens. Bioelectron.*, 2006, **21**, 1178.
42. X.-T. Mo, Y.-P. Zhou, H. Lei and L. Deng, *Enzyme Microb. Technol.*, 2002, **30**, 583.
43. S. Tombelli, M. Mascini, C. Sacco and A. P. F. Turner, *Anal. Chim. Acta*, 2000, **418**, 1.
44. R. Miranda-Castro, P. de-los-Santos-Álvarez, M. J. Lobo-Castano, A. J. Miranda-Ordieres and P. Tunon-Blanco, *Electroanalysis*, 2009, **21**, 267.
45. F. Farabullini, F. Lucarelli, I. Palchetti, G. Marrazza and M. Mascini, *Biosens. Bioelectron.*, 2007, **22**, 1544.
46. J. Wang, *Anal.Chim. Acta*, 2002, **469**, 63.
47. A. Lermo, S. Campoy, J. Barbe, S. Hernandez, S. Alegret and M. I. Pividori, *Biosens. Bioelectron.*, 2007, **22**, 2010.
48. A. Lermo, E. Zacco, J. Barak, M. Delwiche, S. Campoy, J. Barbe, S. Alegret and M. I. Pividori, *Biosensors and Bioelectronics*, 2008, **23**, 1805.
49. S. Liebana, A. Lermo, S. Campoy, J. Barbe, S. Alegret and M. I. Pividori, *Anal. Chem.*, 2009, **81**, 5812.
50. O. A. Loaiza, S. Campuzano, M. Pedrero, P. Garcia and J. M. Pingarron, *Analyst*, 2009, **134**, 34.
51. M. J. LaGier, C. A. Scholin, J. W. Fell, J. Wang and K. D. Goodwin, *Marine Pollut. Bull.*, 2005, **50**, 1251.
52. M. J. LaGier, J. W. Fell and K. D. Goodwin, *Marine Pollut. Bull.*, 2007, **54**, 757.
53. K. Metfies, S. Huljic, M. Lange and L. K. Medlin, *Biosens. Bioelectron.*, 2005, **20**, 1349.
54. A. J. Baeumner, R. N. Cohen, V. Miksic and J. Min, *Biosens. Bioelectron.*, 2003, **18**, 405.
55. M. B. Esch, A. J. Baeumner and R. A. Durst, *Anal. Chem.*, 2001, **73**, 3162.
56. S. Kwakye, V. N. Goral and A. J. Baeumner, *Biosens. Bioelectron.*, 2006, **2**, 2217.
57. S. Kwakye and A. Baeumner, *Anal. Bioanal.Chem.*, 2003, **376**, 1062.
58. N. V. Zaytseva, R. A. Montagna and A. J. Baeumner, *Anal. Chem.*, 2005, **77**, 7520.

59. B. Elsholz, R. Worl, L. Blohm, J. Albers, H. Feucht, T. Grunwald, B. Jurgen, T. Schweder and R. Hintsche, *Anal. Chem.*, 2006, **78**, 4794.
60. B. P. Nelson, M. R. Liles, K. B. Frederick, R. M. Corn and R. M. Goodman, *Environ. Microbiol.*, 2002, **4**, 735.
61. G. Feriotto, M. Borgatti, C. Mischiati, N. Bianchi and R. Gambari, *J. Agric. Food Chem.*, 2002, **50**, 955.
62. I. Mannelli, M. Minunni, S. Tombelli and M. Mascini, *Biosens. Bioelectron.*, 2003, **18**, 129.
63. B. Meric, K. Kerman, G. Marrazza, I. Palchetti, M. Mascini and M. Ozsoz, *Food Control*, 2004, **15**, 621.
64. M. L. Del Giallo, F. Lucarelli, E. Cosulich, E. Pistarino, B. Santamaria, G. Marrazza and M. Mascini, *Anal. Chem.*, 2005, **77**, 6324.
65. M. U. Ahmed, Q. Hasan, M. M. Hossain, M. Saito and E. Tamiya, *Food Control*, 2010, **21**, 599.
66. R. A. Reynolds III, C. A. Mirkin and R. L. Letsinger, *Pure Appl. Chem.*, 2000, **72**, 229.

CHAPTER 4
# Aptamer-based Biosensor for Environmental Monitoring

LAKSHMI N. CELLA, WILFRED CHEN AND
ASHOK MULCHANDANI

Department of Chemical and Environmental Engineering and Cell, Molecular and Developmental Biology Graduate Program, University of California, Riverside, CA 92521, USA

## 4.1 Introduction to Aptamers

Since the discovery of nucleic acids (DNA and RNA), there has been a constant flux of knowledge with respect to their structure and functions. One such finding was the identification by two separate research groups in 1990, through *in vitro* selection, of a class of short RNA molecules that could specifically bind to nucleic acid binding proteins and organic dyes used in affinity columns.[1,2] DNA molecules displaying similar binding capability to organic dyes followed shortly.[3] These molecules were named *aptamers*, based on the Latin word *aptus*, meaning "fit".[4] The complex three-dimensional shapes of the aptamer molecules give them the ability to bind selectively to the target molecules. In a sufficiently diverse, degenerate libraries of $\geq 10^{15}$ molecules, a probable DNA/RNA molecule that can bind to any given target molecule could be found.[2] The *in vitro* selection protocol for isolating aptamers from such diverse libraries is called *systematic evolution of ligands by exponential enrichment (SELEX)* (Figure 4.1). This selection protocol has been remodeled to suit a variety of targets and in recent years even reduced to an automated *in vitro* process leading to high-throughput selection.

---

Nucleic Acid Biosensors for Environmental Pollution Monitoring
Edited by Marco Mascini and Ilaria Palchetti
© Royal Society of Chemistry 2011
Published by the Royal Society of Chemistry, www.rsc.org

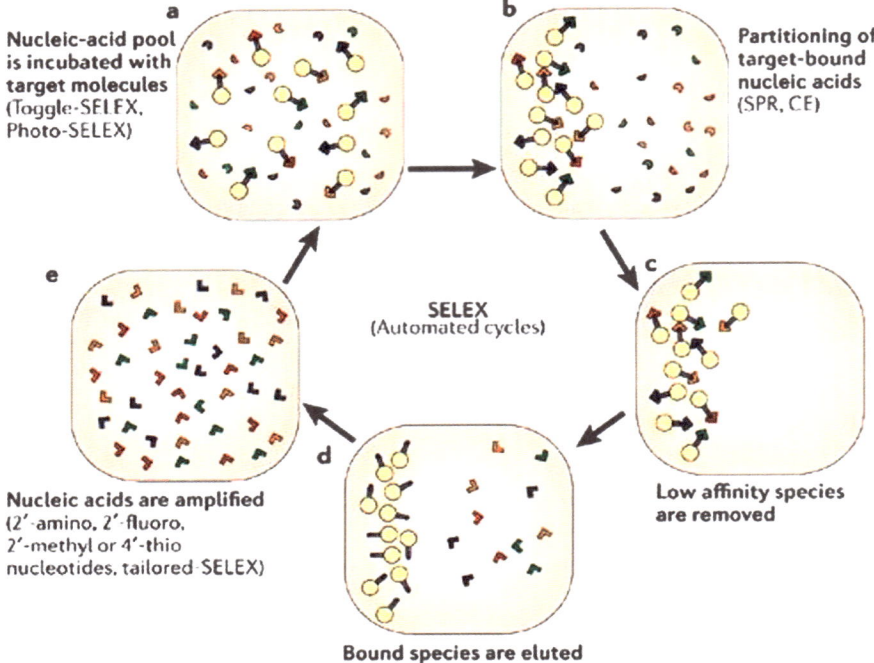

**Figure 4.1** Schematic of SELEX process.[62]

Since their discovery, aptamers have captured the attention of researchers. Today aptamers are available for a wide variety of target molecules ranging from metal ions to whole cells. Several favorable attributes displayed by aptamers—characteristics of high affinity and selectivity, almost equal to or more than that of antibodies, their cost-effective *in vitro* synthesis and stability in a wide range of environments—have led to their employment in sensor platforms for a wide variety of applications. In recent years an in-depth understanding of their structure–function relationship has been attained, leading to many new aptamer-based bioassay systems.[5–7]

## 4.2 Aptamer Properties

Because single-stranded oligonucleotide molecules are highly favorable for forming secondary structures, the number of stable conformations among them can exceed that of a peptide sequence of similar length. This forms the basis for the successful development of aptamers for a variety of target molecules. Most of the aptamers generated display an affinity constant in the range of picomolar to nanomolar. RNA aptamers are naturally more flexible than the DNA molecules and can offer a wider range of conformational diversity. However, the generation of a RNA random pool involves the *in vitro* transcription of a random DNA library in the enrichment process. Susceptibility to degradation by nucleases is another limitation that has restricted the use of RNA in biological fluids. Traditionally the nuclease resistance of RNA aptamers was achieved by introducing modifications into the backbone after selection. The modification at the 2′ position of the ribose sugar remains one of the widely used modifications for increasing the stability of the RNA library.[8,9] However, through these modifications there is a probability of affecting the interactions between the aptamer and target molecule or affecting the aptamer secondary structure, leading to the loss of affinity. As an alternative, in recent years aptamers have been selected from modified RNA libraries bearing suitable modifications, predominantly to the oligonucleotide backbone, and this has led to the generation of nuclease-resistant RNA molecules.[10,11] The SELEX process with the modified libraries is facilitated by the ability of the enzymes employed in the process (polymerases and reverse transcriptases) to recognize the modified nucleotides and incorporate them into the oligonucleotide chain.[12] DNA libraries, and subsequently the selected aptamers in their natural state, are far more robust and display high levels of stability for biological applications. The SELEX process for the DNA aptamers is also further simplified by the elimination of *in vitro* transcription steps. Additional protection for the DNA ligands from exonucleases can be provided by terminal capping, which is a means of immobilizing the aptamer moiety onto a platform in most diagnostic applications. Several different kind of substitutions, at various locations in the nucleotide bases, are used in making of modified oligonucleotide libraries. Apart from alleviating the problem of aptamer stability, some of these modifications increase the diversity of the random library and can suit special selection needs. The modification at the 2′ position of the sugar provides nuclease resistance, whereas modifications at the C-5 position of the pyrimidine bases could be used to attract certain classes of targets through formation of cross-links between the target and the library. One other unique way of further improving the affinity and specificity of aptamers is by performing the selection using a biased library. The biased library is constructed by carrying out error-prone rounds of the polymerase chain reaction (PCR) with the target aptamer as template, increasing the complexity of molecules and displaying species that were not present in the initial random library.

## 4.3 SELEX and its Variants

The SELEX process for selecting high-affinity aptamers from a random pool of molecules involves several steps. The first step is to design the random library; this involves deciding the length of the randomized region, which decides the library size, and fixed regions for library amplification in further rounds of the process. Commonly a library containing 15–75 nucleotides (nt) in the random positions flanked by fixed regions of around 20 nucleotides is chosen. Although the diversity of the randomized library is dependent on the length of random region, the limitations of the solid-phase DNA synthesis technique limits the diversity to $10^{15}$–$10^{16}$. Following the design and synthesis of the random library, it is mixed with target molecules, at a high ratio of target to nucleic acid for the initial rounds, and incubated at a fixed temperature and conditions to allow the binding of the target molecule to the DNA. Subsequently, the bound and unbound molecules are separated by one of the many methods described in the following section. Traditionally the use of nitrocellulose filtration and affinity columns has dominated this process. In recent years several improvements have been made to this process in order to get high-affinity molecules in less time and fewer rounds of selection. The DNA from the bound library is then isolated and amplified by PCR for further rounds of selection. The PCR process in aptamer screening requires increased optimization, as it involves the amplification of the randomized regions of the probable binders. The further rounds of selection/enrichment are carried out with stringent conditions to get high-affinity binders. A round of negative selection or counter-selection is of great importance to remove the nonspecific binders.

In search of efficient separation of bound and unbound libraries and to suit the needs of a particular application, a number of SELEX variants have emerged. A broad classification of the variants will involve methods (1) where the target is immobilized on a support and allowed to interact with the random DNA/RNA library followed by separation of the unbound library and recovery of the bound library from the target by employing harsh conditions; (2) where the interaction between the target and the library occurs in free solution, followed by the separation of the bound and unbound libraries; and (3) specialized variants (Table 4.1).

### 4.3.1 Bound-state SELEX Variants

One fundamental type in this category of SELEX variants involves the use of affinity columns. Affinity surfaces with immobilized target molecules when

Table 4.1 SELEX variants.

| Solution-phase variants | Bound-phase variants | Special variants |
| --- | --- | --- |
| Nitrocellulose membrane filtration | Affinity columns | Toggle SELEX |
| Column matrices | Magnetic separation | Tailored SELEX |
| CE-SELEX | Affinity titer plates | Photo SELEX |
| | SPR-SELEX | |

exposed to random RNA/DNA libraries can interact with certain molecules, thus leading to discovery of new lead compounds. One of the first set of RNA aptamers developed for organic dyes employed affinity columns.[2] The organic dyes were cross-linked with agarose beads and packed in the columns. For selection, a RNA library with 100 nt random sequences flanked by fixed regions was applied to the affinity column. The unbound library was washed away and bound molecules were eluted off the column using high salt concentrations. High-affinity aptamers were obtained after five rounds of SELEX.

As an alternative to packed columns, target molecules have been immobilized on functionalized magnetic particles and the bound and unbound library separated by magnetic separation. An RNA aptamer for Panama influenza virus was isolated in this fashion.[12] The protocol consisted of (1) preparing the target functionalized magnetic beads by coating the whole virus onto epoxy affinity polystyrene beads followed by blocking of nonspecific sites on the beads by bovine serum albumen (BSA); (2) incubation of the RNA random library with the BSA-coated (without the virus) magnetic beads for 10 min to remove any BSA binders; (3) application of the unbound library, recovered after magnetic separation of the beads in step 2 to magnetic beads coated with same subtype A/Aichi virus as a negative selection round; (4) incubation of the unbound library recovered in step 3 with the magnetic beads coated with the Panama influenza A virus; and (5) recovering the library bound to the target functionalized magnetic beads and treating with 7 M urea to isolate the library. The process was repeated for 10 cycles to get high-affinity RNA aptamers. This SELEX variant is less time consuming as it alleviates filtration, centrifugation and other intermediary stages employed for purification of aptamers. Affinity microtitre plates are a commonly employed alternate matrix for immobilization of target molecules.[13] In some studies proteins immobilized by employing tags fused at their N-termini or C-termini have been used. One such example is the RNA aptamer developed for ribosomal protein L22 fused with glutathione S-transferase (GST) tag.[14]

*Surface plasmon resonance SELEX (SPR-SELEX)* involves SPR spectroscopy, a technique for monitoring and studying the interaction between two molecules in real time. It has been used to study interaction of several molecules, ranging from small peptides to phospholipids vesicles. By employing SPR technology for selection of aptamers, not only can we achieve efficient separation of unbound and bound library molecules but also selectively isolate the bound molecules displaying the targeted affinity and thus dramatically reduce the number of selection cycles. Using this methodology, an RNA aptamer was selected against human influenza A (H3N2) virus.[15] The SELEX process involved injecting a random RNA library onto the CM4 chip coated with the virus. The unbound library was removed and the bound library was collected by flowing the binding buffer. Because the method involves immobilizing the target on a support matrix, there is potential for steric hindrance of possible aptamer binding sites on the target and selection of matrix binders.

### 4.3.2 Solution-based SELEX Variants

In this set of SELEX variants the interaction of the random library molecules and the target molecules occurs in free solution, followed by a separate stage for the separation of bound and unbound library. The solution-based selection alleviates steric hindrance in the interaction between the DNA/RNA sequences and target molecules encountered when the target is immobilized. Some of the earliest successful trials of *in vitro* selection involved solution-based interaction and the use of nitrocellulose filtration and column matrices—major classes of solution-based SELEX variants—for separation. Tuerk and Gold[1] were the first to demonstrate the use of nitrocellulose membrane for the separation of bound and unbound library of RNAs for the selection of T4 DNA polymerase (gp43) RNA aptamer. The average efficiency of RNA recovery from the membrane filter was about 80%. After the first trial by pioneers of the SELEX process, several researchers have used the nitrocellulose filtration technique to obtain high-affinity binders for several targets. Since the nitrocellulose membrane filtration is very efficient for separating RNA from its protein complex, this technique has mostly been developed for RNA aptamers rather than DNA aptamers. Affinity gel columns have been used as an alternative to filtration. For example, a DNA aptamer against cellobiose was identified using a cellulose column.[16] Similarly, an RNA aptamer against Zn ions was selected using a chelating sepharose high-affinity column.[17] The bound RNA molecules were eluted by buffer containing EDTA and, at the end of seven rounds of selection, 73% of the selected pool displayed affinity towards Zn.

*Capillary electrophoresis SELEX (CE-SELEX)* has been widely used in recent times because of its speed and higher resolution capacity. A typical selection process (Figure 4.2) involves fewer than 10 cycles for obtaining high-affinity aptamers. Since the process occurs in solution, potential binding sites on the target molecule are not blocked, removing the selection bias induced in the solid-phase-bound SELEX variants. CE-SELEX involves the separation of bound library from the unbound molecules utilizing the mobility shift occurring as a result of complex formation. A typical process starts with determining the mobility of the target molecule and the free library in the capillary. Based on this mobility pattern the window for efficient collection of library–target complex molecules is established. High-affinity aptamers for IgE and PA toxin component of anthrax toxin were selected within four and six rounds of selection, respectively, using this process.[18,21]

### 4.3.3 Specialized Variants

This class of SELEX techniques, which may be conducted in either solution phase or bound phase, involves steps or complexities additional to the conventional SELEX process directed for particular applications. These less tedious directed approaches greatly reduce the time of selection for high-affinity aptamers. *"Toggle" SELEX*, for example, is a directed approach involving the selection of aptamers for a group of closely related molecules by alternating the target molecules between SELEX cycles. White *et al.* developed this process to

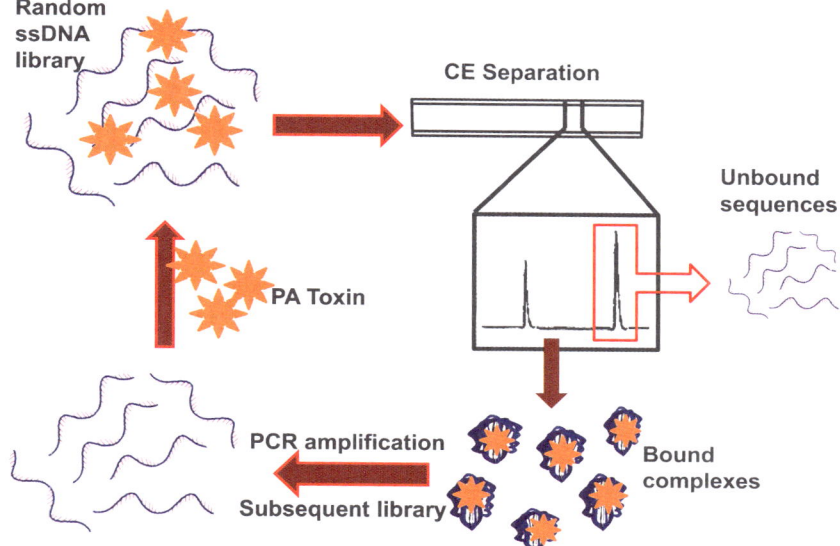

**Figure 4.2** Schematic representation of CE SELEX process.

identify a thrombin aptamer with cross-reactivity for porcine and human forms of thrombin. The selection of targets with known degree of cross-reactivity is desired during preclinical evaluation of potential therapeutic targets.[19] Similarly, *"tailored" SELEX*, a process aiming at developing the shorter aptamer sequences, involves the ligation and cleavage of primer sites at every enrichment round.[20] *Photochemical SELEX (photo-SELEX)* utilizes light-sensitive nucleotides incorporated in the random libraries. Upon activation by light absorption these modified nucleotides can form covalent links to the specific target molecules, providing more specificity to the selection process.[21]

## 4.4 Aptamer-based Biosensors for Environmental Monitoring

Today there is an ever-increasing need to monitor our environment—food, air, water—as the chances of finding agents that can affect human and animal life are huge. Some of these agents can have mild to severe short-term or long-term effects and some of them can even have deadly effects and lead to widespread havoc. In view of the increasing threat of terrorist organizations and rogue nations using biowarfare agents, biosecurity has received a lot of attention from governments.

The agents that need monitoring in the environment can be broadly classified as small organic and inorganic pollutants, pharmaceuticals and personal care products (PPCPs), toxins of microbial origin, and pathogens. Although there has been a lot of effort in developing techniques for environmental monitoring of various agents, there is still a great need for portable, field deployable and highly robust technologies.[22] Traditionally, most of these monitoring techniques and devices utilize antibodies to capture the target molecule and determine its concentration in the sample. Because of the expensive animal models required for their synthesis, unavailability against nonimmunogenic compounds and instability on exposure to varying environments, antibodies are not an ideal choice for environmental monitoring devices, bringing RNA or DNA aptamers into the limelight as an alternative. Apart from having the same or even higher sensitivity and selectivity as antibodies for a target molecule, aptamers can be produced on a large scale utilizing less expensive *in vitro* synthesis, exhibit greater environmental stability and can be generated against a whole range of target molecules. The following section gives a few examples of aptamer-based biosensors for environmental monitoring reported in the literature.

## 4.4.1 Detection of Organic and Inorganic Pollutants

The spreading of industrial wastewater and treated sewage water over agricultural lands and their use for irrigational purposes is an attractive means of disposal and to some extent serves as an inexpensive means of irrigation in areas of water shortage. However, these sources of water are increasingly contaminated by persistent organic (phenanthrene, 1,2,4-tricholorobenzene) and inorganic (heavy metals, *e.g.*, Cd, Pb, Cu, Zn) pollutants, leading to increased environmental concern and investigation of their long-term ecological effects. Several analytical techniques have been developed for sensitive detection of these pollutants including aptamer-based sensors.

Sensors for monitoring inorganic metal ions have wide applications, ranging from environmental monitoring to toxicological studies. Rajendran *et al.*[23] reported a novel method for selection of fluorescent aptamer beacons that can be used for metal ion detection. In their technique the random library labeled with a fluorophore was mixed with 5′ dabcyl-labeled capture oligo (short complementary oligonucleotide) and introduced into affinity column containing receptor for capture oligo. Upon addition of $Zn^{2+}$ the random library molecules displaying affinity towards $Zn^{2+}$ bind to it and undergo conformational changes leading to release from the column. Thus, this method not only selects aptamers of high affinity but also leads to fluorescent signaling without further modifications. The assay had a limit of detection of 5 μM, dynamic range up to 2 mM of $Zn^{2+}$ and demonstrated high selectivity against other metal ions except cadmium.

Several aptamer-based sensors have been developed for detection of small organic molecules and these novel platforms can been adapted for efficient detection of similar hazardous environmental pollutants. For example, several aptamer-based biosensors have been reported for a small molecule (cocaine) using a 39-mer DNA anti-cocaine aptamer with $K_d$ of 5 μM.[24] Stojanovic *et al.*[24] applied the aptamer in a Förster resonance energy transfer (FRET)-based biosensor by splitting the aptamer into two subunits based on its predicted

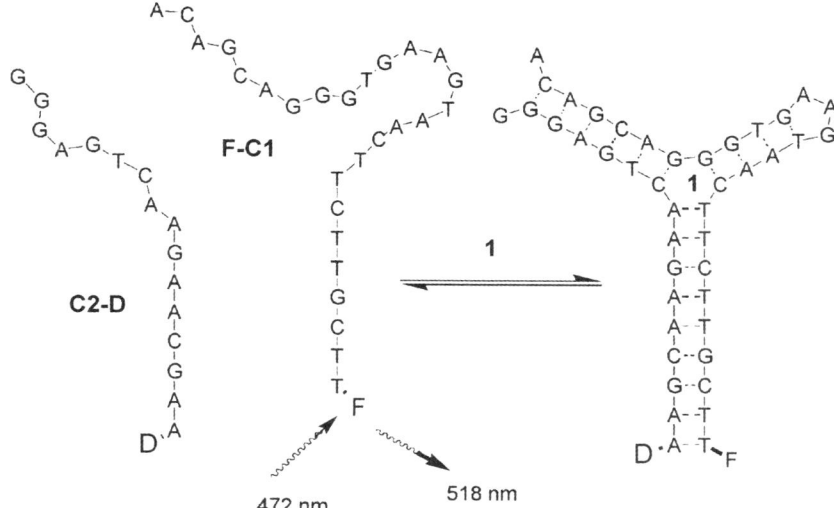

**Figure 4.3** Detection of cocaine based on the FRET occurring between the two subunits of the cocaine aptamer with subunit F-C1 labeled with 6-FAM and subunit C2-D labeled with a quencher, dabcyl. The FRET assay system is on until the introduction of cocaine into the reaction, which brings together the two subunits, quenching the fluorescence.[24]

secondary structure. As shown in Figure 4.3, one of the subunits was labeled with donor 5'-6-carboxy-fluorescein (6 FAM) and other with 3'-dabcyl quencher. In the nonhybridized state there was little energy quenching, because of the large separation distance between the fluorescent dye and quencher. In the presence of cocaine and at optimized concentrations of the subunit oligos, the two subunits are assembled, quenching the fluorescence. This sensor reliably measured cocaine in the range 10–1250 µM.

Liu et al.[25] reported a colorimetric biosensor for cocaine based on gold nanoparticles and cocaine aptamers. The biosensor consists of three components: gold nanoparticles (13 nm diameter) functionalized with 3'-thiol-modifies DNA (3' $Coap_{Au}$), 5'-thiol-modified DNA (5' $Coap_{Au}$) and a linker DNA ($Linker_{Coap}$) molecule (Figure 4.4). 3' $Coap_{Au}$ and 5' $Coap_{Au}$ were assembled on the linker molecule to form aggregates. Apart from the cocaine aptamer sequence, the linker molecule has two other components that hybridize with 3' $Coap_{Au}$ and 5' $Coap_{Au}$. A fragment of 5' $Coap_{Au}$ interacts with the cocaine aptamer as well. Upon introduction of cocaine, the aptamer changes conformation and displaces the 5' $Coap_{Au}$ fragment from the aggregates, hence dissolving them and resulting in color change which can be easily quantified. Using this technique cocaine can be measured in the concentration range 50–500 µM. Li et al.[26] developed an electrogenerated chemiluminescence (ECL) aptamer-based biosensor with ruthenium complex serving as ECL label. In this sensor $Ru(bpy)_2(dcbpy)NHS$-labeled aptamer is immobilized onto a gold

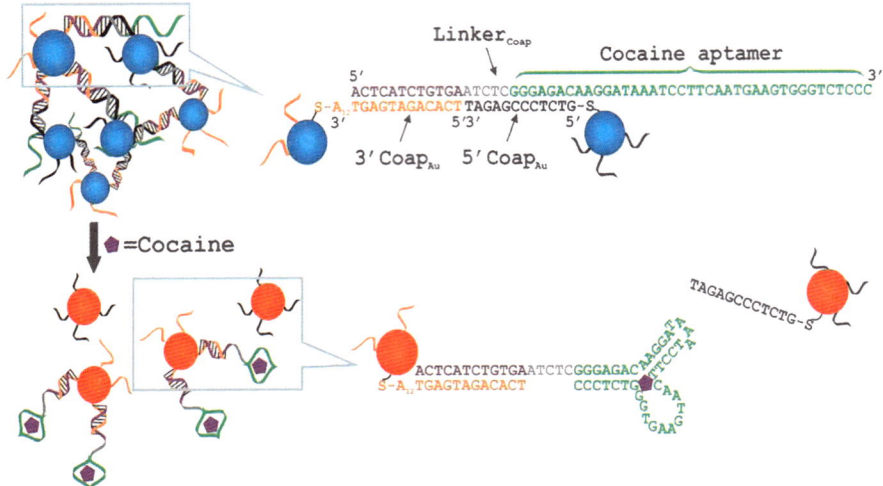

**Figure 4.4** A novel technique for colorimetric detection of cocaine utilizing nanoparticles. The assay solution in the absence of cocaine is purple in color due to the nanoparticle aggregates formed by the linker molecule, containing (1) a region that hybridizes with gold nanoparticle attached 3′ Coap$_{Au}$ oligo, (2) a region that hybridizes with gold nanoparticle functionalized 5′ Coap$_{Au}$ oligo, and (3) the cocaine aptamer itself. Upon addition of cocaine the aptamer folds in to bind to the target and releases the 5′ Coap$_{Au}$ oligo, dissolving the aggregates, and the color of the solution changes from purple to red.[25]

electrode via thiol interactions. In the absence of cocaine the aptamer remains partially unfolded, resulting in small ECL signals which increase in presence of cocaine as the tag is brought close to the electrode. This sensor displayed increased sensitivity by detecting cocaine in a concentration range of 5 nM–30 µM. Other salient features of the ECL aptamer sensor include its reusability and longer shelf life.

### 4.4.2 Detection of Drugs, EDCs, and PPCPs

Antibiotic drugs, endocrine-disrupting chemicals (EDCs) and PPCPs are emerging environmental issues attracting increased attention and concern worldwide from the scientific community, the media and the general public. In recent years, chemicals such as tetracyclines, estrogenic compounds (nonylphenol, β-estradiol, estrone, α-estradiol), bisphenol A and some pesticides have increasingly been found in treated domestic wastewater effluents.[27–31] These compounds are of great concern because of their potential to cause health hazards by altering the normal endocrine function and physiological status of marine ecosystems and humans.[32–34]

Tetracyclines (tetracycline and oxytetracycline) are commonly used as growth promoters for animals, and residual amounts of these compounds have been detected in slaughtered animals and animal products (milk, eggs and honey). Tetracycline is hepatotoxic and severely affects pregnant women and newborn infants. Increasingly, conjugated estrogens used in the treatment of cancer, hormonal imbalance, osteoporosis, and other ailments are also contributing to the increasing source of estrogen pollution.[35] The use of contraceptives and other pharmaceuticals that are estrogenic are also causing severe long-term problems due to their high persistence and biological activity in the environment. Many of these known estrogenic compounds end up in the aquatic environment via sewage, the discharge of municipal and/or industrial effluents, and agricultural run-off. Concentrations up to 80 ng/L of estrogenic compounds have been detected in domestic effluent samples;[36] hence the increased effort to develop sensitive but robust on-site detection techniques for these compounds.

Kim et al.[37] developed high-affinity 76-mer ssDNA aptamers that detected tetracycline[38] (TET, $K_d$ 63.6 nM) and oxytetracycline (OTC, $K_d$ 11 nM). In order to detect these compounds they utilized cyclic voltammetry (CV) and square wave voltammetry (SWV) on a gold modified electrode with immobilized TET aptamer through avidin–biotin interactions (Figure 4.5). The SWV analysis was conducted for a series of tetracycline concentrations (1 nM to 100 μM) with 10 nM as the limit of detection, similar to HPLC, and a dynamic range up to 10 μM. The sensor was specific in the presence of related compounds (oxytetracycline and doxicycline). In case of the OTC sensor the aptamer was attached to an interdigitated gold electrode through simple thiol interactions. The dynamic range for this sensor was 1–100 nM and the limit of detection was 5 nM. The sensor can be reused by treatment with NaCl and displayed the required specificity in presence of closely related compounds.

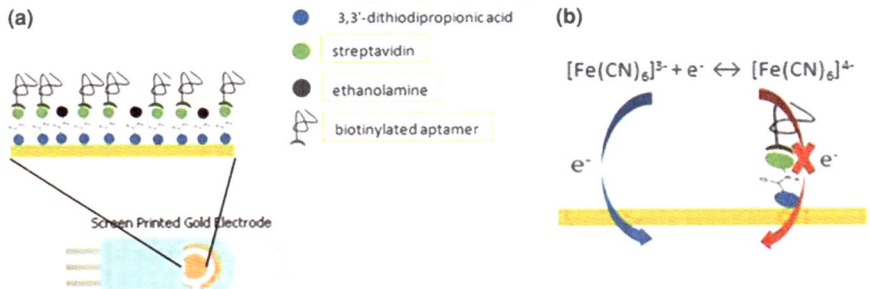

Figure 4.5 (a) Schematic of the biosensor built for detection of tetracycline by immobilization of the biotin-labeled aptamer on the streptavidin-coated gold electrode. (b) Schematic of the detection scheme. In this electrochemical system there is a free flow of electrons between the solution and the electrode, which is interfered with upon introduction of analyte tetracycline due to the aptamer–analyte complex formation.[38]

Neomycin B, an antibiotic, is extremely nephrotoxic; beyond permissible levels. Alvarez et al.[39] used a 2'-O-methylated-modified RNA aptamer ($K_d$ 5 µM) combined with SPR to sensitively detect this antibiotic. The SPR chip was first modified with neomycin B, followed by flowing in of varying concentrations of neomycin B and a fixed concentration of aptamer, leading to competition for binding to the aptamer. A limit of detection of 5 nM and range of quantification between 10 nM and 100 nM was reported.

An electrochemical aptasensor for detection of 17β-estradiol was developed by Kim et al.[40] They selected a 76-mer ssDNA aptamer with an affinity constant of 0.13 µM and immobilized on a streptavidin-modified gold electrode. As in their detection of tetracyclines, they utilized CV and SWV and found a linear range of detection of the compound in the concentration range 1–10 µM.

### 4.4.3 Detection of Toxins

Enforcing biosecurity requires vigilant environmental monitoring for deadly pathogens and toxins introduced for malicious or criminal purposes. There is an ever-increasing threat to civilian and military targets by rogue nations and terrorist organizations. In particular, there is an increasing probability of using weapons of mass destruction such as nuclear, chemical and biological weapons. Biowarfare agents—toxins, pathogens—are preferred weapons of mass destruction as they can be easily produced and disseminated, cause high mortality and require a high degree of planning and preparedness to protect the public in case of their release into environment. Many of the available biothreat sensing devices use antibodies as molecular recognition elements; however, this makes the devices less robust and difficult for onsite deployment. In recent years several aptamers have been generated against deadly toxins and also whole-cell pathogens and highly sensitive sensors have been constructed using them.

Ricin, a toxin from the castor oil plant *Ricinus communis*, is a class II ribosome inactivating protein and displays great potential for being used as a bioweapon.[41,42] There have been several attempts to make high-affinity aptamers against the ricin toxin. The earliest attempt was by Hesselberth et al. to develop an RNA aptamer against ricin A-chain.[43] Using the nitrocellulose filtration SELEX technique, this group generated an 81-mer RNA aptamer ($K_d$ 7.4 nM) and employed it in an array platform along with lysozyme RNA aptamer and IgE and thrombin DNA aptamers for multiplex determination of protein concentrations.[44] They labeled a streptavidin-coated glass slide with 5' biotin-labeled aptamers at specific locations. The proteins, either individually or in a mixture, were labeled with cy5 fluorescent dye or introduced onto the array suspended in optimized PBS buffer containing 5 mM $MgCl_2$ and 0.1% Tween 20 to promote binding. The proteins bound specifically to the individual aptamers with no cross-reactivity. The array sensor could detect as low as 0.5 nM ricin with high selectivity. Utilizing the same anti-ricin aptamer, Haes et al.[45] reported detection of 500 pM ricin using affinity probe capillary electrophoresis. Apart from the high sensitivity they also demonstrated detection in the presence of contaminating matrix elements such as BSA, casein and RNA nucleases. The technique is similar to CE-SELEX, and involved reaction of fluorescently-labeled aptamer with ricin followed by monitoring the ricin-bound and unbound aptamer fluorescence downstream of the capillary electrophoresis using a laser-induced fluorescence detector. Tang et al.[46] generated three high-affinity ricin-binding 40-mer DNA aptamers using CE-SELEX (C1, with $K_d$ 80 nM; C5, with $K_d$ 58 nM) and affinity column SELEX process (A3, with $K_d$ 105 nM). However, these have not yet been adapted for a sensor platform.

Abrin is another class II ribosome inactivating toxin from the legume *Abrus precatorius* which, like ricin, is extremely toxic to the eukaryotic cells and has a great potential as a biowarfare agent. Using affinity chromatography SELEX, Tang et al.[47] generated a high-affinity 35-mer ssDNA aptamers (TA6, $K_d$ 28 nM) against the abrin toxin. This aptamer was applied for arbin detection using a new light-switching reagent $[Ru(phen)_2(dppz)]^{2+}$.[48] This reagent has a unique property of having no luminescence in aqueous solution but emitting strong luminescence upon intercalating to the duplex nucleic acid. When the ssDNA aptamer is not bound to the target, it forms a secondary structure and $[Ru(phen)_2(dppz)]^{2+}$ can intercalate to emit luminescence. Upon addition of target molecule the aptamer changes its conformation to bind/accommodate the target ligand and displaces the light-switching reagent, leading to change in the luminescence value. Among their high-affinity aptamers, aptamer TA6 showed the best signaling performance and abrin toxin concentration as low as 1 nM could be detected. The sensor displayed a linear response from 0 to 400 nM of abrin concentration and high selectivity for the toxin even in presence of interfering proteins.

Anthrax is a disease caused by the spore-forming bacterium *Bacillus anthracis*. Upon entry into a susceptible host it produces a three-component toxin: protective antigen (PA), edema factor (EF) and lethal factor (LF). This can lead to death of the infected individual in a very short time, leaving very little time for treatment.[49] Early detection of the disease is also challenging because of its fairly nonspecific symptoms, as evidenced by the identification of infection before death in only one of five casualties in the 2001 terrorist attacks in the United States, underlining the need for sensitive detection technology for the anthrax.[50]

Cella et al.[51] developed a chemiresistive aptasensor for the detection of PA toxin based on single-walled carbon nanotubes (SWNTs). They generated a high-affinity 40-mer ssDNA aptamer ($K_d$ 112 nM) against PA toxin using the CE-SELEX process in six rounds of enrichment. The aptamer was immobilized onto AC dielectrophoretically aligned SWNTs across a 3 μm gap microfabricated gold electrode through a linker molecule 1-pyrenebutanoic acid succinimidyl ester[52] (Figure 4.6). The binding of PA toxin to the aptamer led to a change in the conductance of the SWNT channel, and upon quantification a wide dynamic range up to 800 nM with a linear range up to 400 nM and a detection limit of 16 nM was established for the sensor.

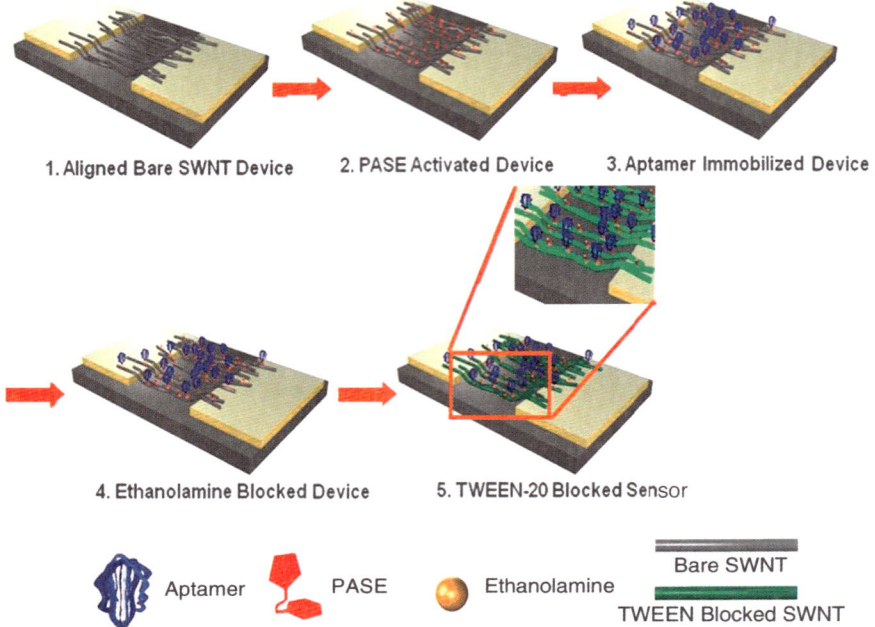

**Figure 4.6** Schematic description of the steps involved in the fabrication of biosensor for detection of PA toxin.[51]

Botulinum neurotoxin (BoNT), produced by the bacterium *Clostridium botulinum* found in improperly handled meat and dairy products, is one of the most toxic substances known and can lead to widespread casualties within days.[53] Among the seven defined serotypes of BoNT (A–G), type A is the most frequent cause of poisoning and human botulism.[54] Wei and Ho[55] used a 76-mer ssDNA aptamer reported by Tok and Fisher[56] against aldehyde-inactivated BoNT/A ($K_d$ 3 nM) and combined its conformational change upon binding to BoNT/A with an electrochemical method and enzymatic amplification. As shown in Figure 4.7, the electrochemical sensor consists of streptavidin-dendrimer-modified polypyrrole substrate to which the aptamer was anchored through the biotin tag. The other end of aptamer was tagged with fluorescein (reporting label). For signal amplification, anti-fluorescein antibody conjugated to horseradish peroxidase (HRP) was used. When the aptamer is in the bound state it displays a more relaxed conformation and the reporter tag is more accessible to the antibody.

Once the antibody had interacted with the fluorescein tag, 3,3′,5,5′ tetramethylbenzidine (TMB/$H_2O_2$) was added and a reductive potential was applied to the polypyrrole electrode to oxidize the HRP mediated by TMB and thus amplifying the signal. In the absence of the BoNT/A toxin the aptamer forms a tightly coiled structure and the reporter tag is inaccessible for the antibody

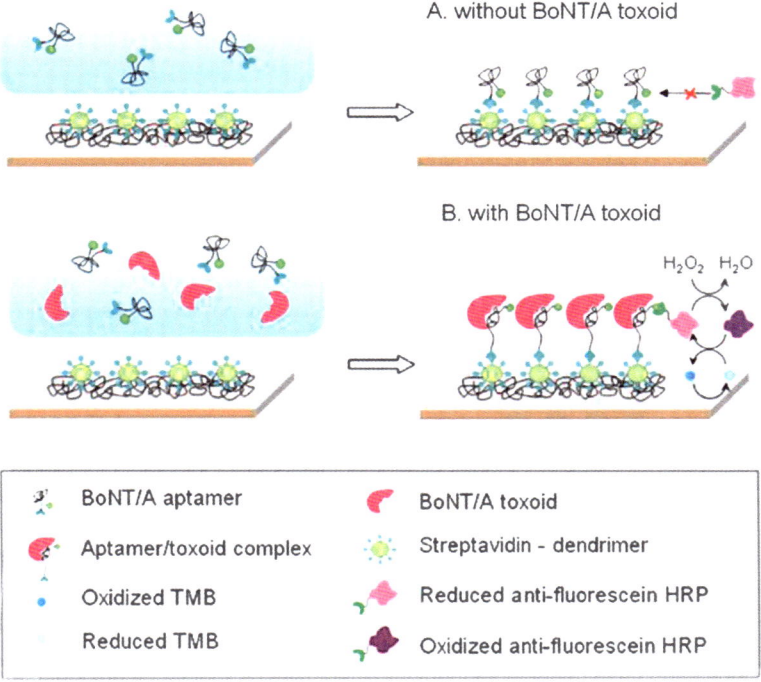

**Figure 4.7** Electrochemical detection of botulinum neurotoxin. Here the aptamer is immobilized on streptavidin-dendrimer-interfaced polypyrrole substrate through its biotin tag. The aptamer is also tagged with a reporting tag, fluorescein, which in the presence of toxin is more accessible to reaction with a HRP-labelled antibody. The signal amplification mediated through TMB occurs only in the presence of toxin as the aptamer is in the relaxed state.[55]

interaction. With this technique the dynamic range of detection was 100 ng/ml to 40 ng/ml, and a limit of detection of 40 ng/ml within 24 h was established.

### 4.4.4 Detection of Pathogens

The rapid detection and monitoring of pathogens, bacteria, viruses and spores in the environment is a persisting need for the environment, food and personal safety. The presence of *Escherichia coli* in food and water can cause serious effects to health, especially for infants, elderly people and immunocompromised patients. Traditionally foodborne pathogens are detected using culture-based techniques, which are labor intensive and time consuming. So *et al.*[57] generated an ssRNA aptamer against *E. coli* strain DH5α and immobilized it on an SWNT-FET (SWNT field-effect transistor) to measure most probable number (MPN), a method used by microbiologists to estimate the number of microorganisms. The SWNT-FET was fabricated by a patterned growth technique with a 5 μM distance between the source and drain electrode and ssRNA aptamers immobilized. Before applying the *E. coli* solution to the sensor they were fixed with 2.5% glutaraldehyde in phosphate buffer and dehydrated with alcohol. The >50% change in conductance of SWNT-FET channel upon addition of 3 μL sample was considered a positive signal. However, the estimates of the titer from the SWNT-FET device and the culture method differed by a factor of two, which was attributed to the smaller size of the device and smaller sample volume.

One other whole-organism pathogen which requires prompt detection and efficient control mechanisms to prevent widespread pandemics and epidemics is the influenza virus. This virus expresses two major surface glycoproteins, hemagglutinin (HA) and neuraminidase (NA), which can be targeted to generate vaccines. However, with every generation these proteins undergo mutations, leading to new subtypes and making vaccine generation much more difficult. In their first series of experiments Kumar *et al.*[15] generated high-affinity RNA aptamers against purified HA of human influenza A virus subtype H3N2 using SPR-SELEX. The aptamer chosen after 10 cycles of selection had a $K_d$ of 188 nM and was observed to inhibit the HA-mediated membrane fusion, underlining its anti-influenza therapeutic value. In their next approach the same group generated a RNA aptamer for HA of influenza B virus in nine cycles of SPR-SELEX. One aptamer at the end of selection had a $K_d$ of 720 pM and could discriminate between influenza A and B virus. Jeon *et al.*,[58] on the other hand, developed a ssDNA aptamer against specific domain of HA protein and showed that it prevented infection by blocking the binding of virus to the cell surface. However, this group does not describe the binding affinities and specificities displayed by the aptamer. These aptamers developed as anti-influenza therapeutics could very be well extended and adapted into suitable technologies or platforms for highly specific molecular diagnostic technologies.

Under unfavorable conditions the anthrax-causing microorganism *Bacillus anthracis* transforms into spores. These are dormant and extremely stable, but can easily be converted back to the vegetative state when they come in contact with host organisms, causing full-blown disease. Bruno *et al.*[58] generated 40-mer ssDNA aptamers against *B. anthracis* spores and used aptamer magnetic electrochemiluminescence assay (AM-ECL) for their sensitive detection. They employed a SELEX variant wherein the unbound random DNA library was separated from the DNA spore complex by centrifugation, and the bound DNA was then heat-eluted for further enrichment. They carried out four rounds of selection in total with different concentration ratios of DNA to spore. At the end of fourth round the aptamers were collected and two copies of them were made: a capture set, capable of attaching to the magnetic beads, and a reporter set tagged with avidin. Initially the capture pool was incubated with the magnetic beads to immobilize them and later with *B. anthracis* spores. Upon thorough washing the reporter set aptamers were added followed by streptavidin-$Ru(bpy)_3^{2+}$. Upon introduction of 0.2 M tripropylamine (TPA) and application of voltage, $Ru(bpy)_3^{2+}$ and TPA undergo reaction and release a photon at 620 nm indicative of the spore concentration. The ECL sensor displayed a wide dynamic range of 10 to $6\times10^6$ with high selectivity. Another research group, Ikanovic et al.,[59] developed a fluorescence assay based on aptamer-quantum dot binding to *B. thuringinesis* spores. This group developed a 60-mer ssDNA aptamer using the centrifugal SELEX process similar to that of Bruno et al. over five rounds. The aptamers were conjugated to fluorescent 655 nm quantum dots (QD). The assay exploits the inherent fluorescent properties of the spores. Binding of these to the aptamer–QD complex leads to a change in fluorescent spectra which can be used for quantification. This sensor has a limit of detection between $10^3$ and $10^4$ and the detection methodology could very well be extended to other similar strains of *Bacillus*.

### 4.4.5 Detection of Nitroaromatic Explosives

In order to insure efficient decontamination of the soil and surrounding environment of past military activity, highly sensitive detection technologies are greatly needed. The remnant explosive materials can remain in the area for a long time. They seep into the groundwater and can give rise to carcinogenic derivatives on undergoing metabolism in the human body.[59,60] Trinitrotoluene (TNT), one of the most widely used explosives worldwide, is of major concern as it is very persistent in the environment and can cause serious health hazards. Detection technologies are required, not only for detection of TNT in actual sites of use but for detection of trace levels in the distant environment. Ehrentreich-Forster et al.[61] developed a robust and high sensitivity aptamer-based fiber-optic sensor. An ssRNA aptamer with 90 random nucleotides was generated against TNT using affinity SELEX after 13 rounds. The aptamer was labeled with a 200 nm fluorescent nanobead by annealing an oligo modified with the fluorescent nanobead using the biotin tag to the additional known specific complementary sequence added to the aptamer. The fluorescently labeled aptamer was then mixed with the analyte sample and the mixture flowed over the surface of a fused silica fiber modified with TNT analog acting as waveguide attached to a photomultiplier detector (Figure 4.8). The detector response was inversely related to the TNT concentration in the sample. Using this technique the group achieved a detection limit in low picomolar ranges, with high selectivity as well.

## 4.5 Future Prospects

Because of their *in vitro* selection and generation, chemical synthesis, modification on demand and stability, aptamers have great potential as next-generation biological recognition elements in affinity-based biosensors for environmental monitoring. Some early indications of this potential have been reported in this chapter. However, the realization of the full potential of aptamer-based sensors will require further research and development in (1) generating aptamers with higher affinity and specificity for improved limits of detection, and detection in complex samples to alleviate or minimize the need

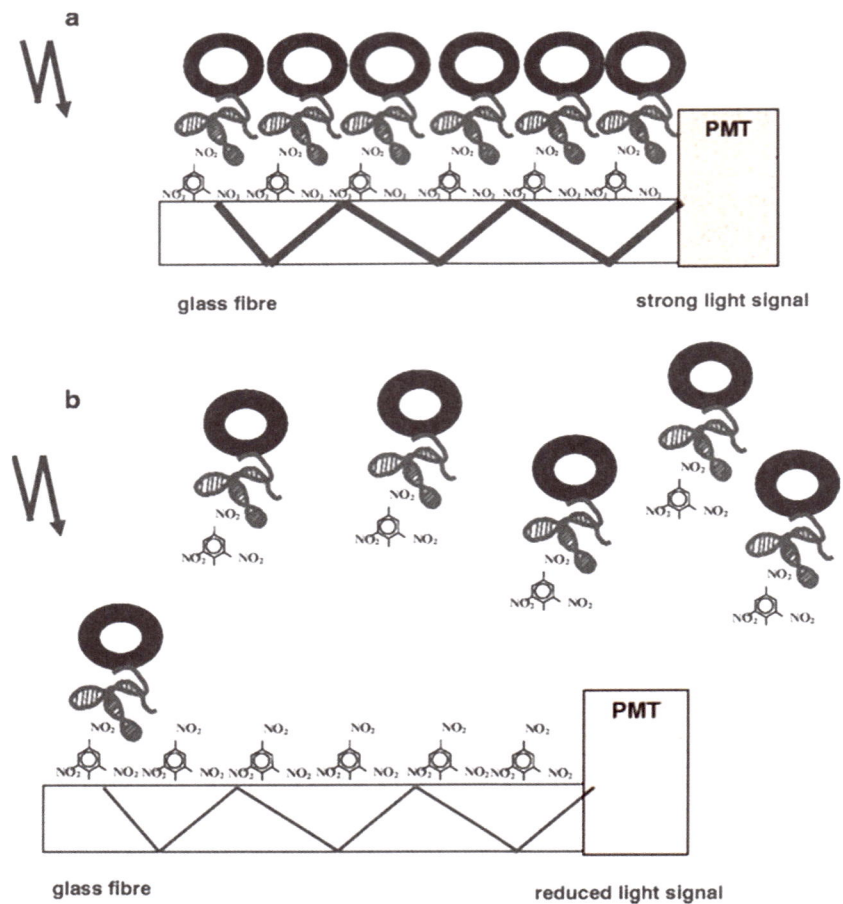

**Figure 4.8** (a) Schematic of the optical sensor, with immobilized TNT with fluorescent bead-labeled aptamer attached to it. (b) The detection of TNT involves the reduction in the light signal reaching the detector through the fiber as a result of displacement of TNT aptamer by the free TNT in solution flowing through the cell.[61]

for sample preparation/preconcentration; (2) development of reagentless, robust, reliable, inline/online, rapid and affordable systems that are field-deployable and autonomous; (3) development of arrays capable of multianalyte sensing; and (4) rigorous evaluation of the sensors in relevant samples.[63]

# References

1. C. Tuerk and L. Gold, *Science*, 1990, **249**, 505.
2. A. D. Ellington and J. W. Szostak, *Nature*, 1990, **346**, 818.

3. A. D. Ellington and J. W. Szostak, *Nature*, 1992, **355**, 850.
4. S. D. Jayasena, *Clin. Chem.*, 1999, **45**, 1628.
5. L. C. Stacey and T. R. Vincent, *Electrophoresis*, 2002, **23**, 1335.
6. W. Itamar and Z. Maya, *Angew. Chem. Int. Ed.*, 2007, **46**, 6408.
7. S. Tombelli, M. Minunni and M. Mascini, *Biosens. Bioelectron.*, 2005, **20**, 2424.
8. J. Hobbs, H. Sternbac, M. Sprinzl and F. Eckstein, *Biochemistry*, 1973, **12**, 5138.
9. W. A. Pieken, D. B. Olsen, F. Benseler, H. Aurup and F. Eckstein, *Science*, 1991, **253**, 314.
10. E. Uhlmann and A. Peyman, *Chem. Rev.*, 1990, **90**, 543.
11. S. Agrawal, *Trends Biotechnol.*, 1996, **14**, 376.
12. A. D. Keefe and S. T. Cload, *Curr. Opin. Chem. Biol.*, 2008, **12**, 448.
13. A. Rhodes, A. Deakin, J. Spaull, B. Coomber, A. Aitken, P. Life and S. Rees, *J. Biol. Chem.*, 2000, **275**, 28555.
14. M. Dobbelstein and T. Shenk, *J. Virol.*, 1995, **69**, 8027.
15. T. S. Misono and P. K. R. Kumar, *Anal. Biochem.*, 2005, **342**, 312.
16. Q. Yang, I. J. Goldstein, H. Y. Mei and D. R. Engelke, *Proc. Natl. Acad. Sci. U. S. A.*, 1998, **95**, 5462.
17. J. Ciesiolka, J. Gorski and M. Yarus, *RNA*, 1995, **1**, 538.
18. S. D. Mendonsa and M. T. Bowser, *Anal. Chem.*, 2004, **76**, 5387.
19. R. White, C. Rusconi, E. Scardino, A. Wolberg, J. Lawson, M. Hoffman and B. Sullenger, *Mol. Therapy*, 2001, **4**, 567.
20. A. Vater, F. Jarosch, K. Buchner and S. Klussmann, *Nucleic Acids Res.*, 2003, **31**, e130.
21. M. C. Golden, B. D. Collins, M. C. Willis and T. H. Koch, *J. Biotechnol.*, 2000, **81**, 167.
22. B. Durodie, *Curr. Opin. Biotechnol.*, 2004, **15**, 264.
23. M. Rajendran and A. D. Ellington, *Anal. Bioanal. Chem.*, 2008, **390**, 1067.
24. M. N. Stojanovic, P. de Prada and D. W. Landry, *J. Am. Chem. Soc.*, 2000, **122**, 11547.
25. J. W. Liu and Y. Lu, *Angew. Chem. Int. Ed.*, 2006, **45**, 90.
26. Y. Li, H. L. Qi, Y. Peng, J. Yang and C. X. Zhang, *Electrochem. Commun.*, 2007, **9**, 2571.
27. C. Baronti, R. Curini, G. D'Ascenzo, A. Di Corcia, A. Gentili and R. Samperi, *Environ. Sci. Technol.*, 2000, **34**, 5059.
28. H. R. Aerni, B. Kobler, B. V. Rutishauser, F. E. Wettstein, R. Fischer, W. Giger, A. Hungerbuhler, M. D. Marazuela, A. Peter, R. Schonenberger, A. C. Vogeli, M. J. F. Suter and R. I. L. Eggen, *Anal. Bioanal. Chem.*, 2004, **378**, 688.
29. M. Peck, R. W. Gibson, A. Kortenkamp and E. M. Hill, *Environ. Toxicol. Chem.*, 2004, **23**, 945.
30. T. Nakari, *Environ. Toxicol.*, 2004, **19**, 207.
31. J. Lintelmann, A. Katayama, N. Kurihara, L. Shore and A. Wenzel, *Pure Appl. Chem.*, 2003, **75**, 631.
32. W. H. Moger, *Can. J. Physiol. Pharmacol.*, 1980, **58**, 1011.

33. B. Hoffmann and A. Landeck, *Anim. Reprod. Sci.*, 1999, **57**, 89.
34. K. E. Tollefsen, J. F. A. Meys, J. Frydenlund and J. Stenersen, *Marine Environ. Res.*, 2002, **54**, 697.
35. M. J. Rosenberg, A. Meyers and V. Roy, *Contraception*, 1999, **60**, 321.
36. C. Desbrow, E. J. Routledge, G. C. Brighty, J. P. Sumpter and M. Waldock, *Environ. Sci. Technol.*, 1998, **32**, 1549.
37. Y. S. Kim, J. H. Niazi and M. B. Gu, *Anal. Chim. Acta*, 2009, **634**, 250.
38. Y.-J. Kim, Y. S. Kim, J. H. Niazi and M. B. Gu, *Bioprocess. Biosyst. Eng.*, 2010, **33**, 31.
39. N. de-los-Santos-Alvarez, M. J. Lobo-Castanon, A. J. Miranda-Ordieres and P. Tunon-Blanco, *Biosens. Bioelectron.*, 2009, **24**, 2547.
40. Y. S. Kim, H. S. Jung, T. Matsuura, H. Y. Lee, T. Kawai and M. B. Gu, *Biosens. Bioelectron.*, 2007, **22**, 2525.
41. R. A. Zilinskas, *JAMA*, 1997, **278**, 418.
42. S. L. Wiener, *Mil. Med.*, 1996, **161**, 251.
43. J. R. Hesselberth, D. Miller, J. Robertus and A. D. Ellington, *J. Biol. Chem.*, 2000, **275**, 4937.
44. E. J. Cho, J. R. Collett, A. E. Szafranska and A. D. Ellington, *Anal. Chim. Acta*, 2006, **564**, 82.
45. A. J. Haes, B. C. Giordano and G. E. Collins, *Anal. Chem.*, 2006, **78**, 3758.
46. J. J. Tang, J. W. Xie, N. S. Shao and Y. Yan, *Electrophoresis*, 2006, **27**, 1303.
47. J. J. Tang, T. Yu, L. Guo, J. W. Xie, N. S. Shao and Z. K. He, *Biosens. Bioelectron.*, 2007, **22**, 2456.
48. C. Hiort, P. Lincoln and B. Norden, *J. Am. Chem. Soc.*, 1993, **115**, 3448.
49. J. C. Milne, D. Furlong, P. C. Hanna, J. S. Wall and R. J. Collier, *J. Biol. Chem.*, 1994, **269**, 20607.
50. J. A. Jernigan, D. S. Stephens, D. A. Ashford, C. Omenaca, M. S. Topiel, M. Galbraith, M. Tapper, T. L. Fisk, S. Zaki, T. Popovic, R. F. Meyer, C. P. Quinn, S. A. Harper, S. K. Fridkin, J. J. Sejvar, C. W. Shepard, M. McConnell, J. Guarner, W. J. Shieh, J. M. Malecki, J. L. Gerberding, J. M. Hughes and B. A. Perkins, *Emerg. Infect. Dis*, 2001, **7**, 933.
51. L. N. Cella, P. Sanchez, W. Zhong, N. V. Myung, W. Chen and A. Mulchandani, *Anal. Chem.*, 2010, **82**, 2042.
52. R. J. Chen, S. Bangsaruntip, K. A. Drouvalakis, N. W. S. Kam, M. Shim, Y. M. Li, W. Kim, P. J. Utz and H. J. Dai, *Proc. Natl. Acad. Sci. U. S. A.*, 2003, **100**, 4984.
53. R. M. Kostrzewa and J. Segura-Aguilar, *Neurotox. Res.*, 2007, **12**, 275.
54. S. S. Arnon, *JAMA*, 2001, **285**, 2081.
55. F. Wei and C. M. Ho, *Analytical and Bioanalytical Chemistry*, 2009, **393**, 1943.
56. J. B. H. Tok and N. O. Fischer, *Chem. Commun.*, 2008, 1883.
57. H. M. So, D. W. Park, E. K. Jeon, Y. H. Kim, B. S. Kim, C. K. Lee, S. Y. Choi, S. C. Kim, H. Chang and J. O. Lee, *Small*, 2008, **4**, 197.
58. S. H. Jeon, B. Kayhan, T. Ben-Yedidia and R. Arnon, *J. Biol. Chem.*, 2004, **279**, 48410.

59. J. A. Styles and M. F. Cross, *Cancer Lett.*, 1983, **20**, 103.
60. B. S. Levine, E. M. Furedi, D. E. Gordon, J. J. Barkley and P. M. Lish, *Fundam. Appl. Toxicol.*, 1990, **15**, 373.
61. E. Ehrentreich-Forster, D. Orgel, A. Krause-Griep, B. Cech, V. A. Erdmann, F. Bier, F. W. Scheller and M. Rimmele, *Anal. Bioanal. Chem.*, 2008, **391**, 1793.
62. D. H. J. Bunka and P. G. Stockley, *Nat. Rev. Microbiol.*, 2006, **4**, 588.
63. N. O. Fischer, T. M. Tarasow and J. B-H. Toh, *Curr. Opinion Chem. Biol.*, 2007, **11**, 316.

CHAPTER 5
# Catalytic Nucleic Acid Biosensors for Environmental Monitoring

NANDINI NAGRAJ AND YI LU

Department of Chemistry, University of Illinois at Urbana-Champaign, Urbana, IL-61801, USA

## 5.1 Discovery of Catalytic Nucleic Acids

The discovery that nucleic acids (NAs) can perform catalytic functions in addition to being genetic information carriers has opened doors to a new paradigm in chemistry and biology. While all biological enzymes had been long regarded as being proteins, discoveries made over the last 30 years have changed this perception. Ribozymes are RNA molecules that catalyze biological reactions and were independently discovered by Cech and Altman in the early 1980s,[1,2] for which they were awarded the Nobel Prize in 1989. This discovery revolutionized the manner in which the roles of RNA molecules were viewed and the hypothesis of an "RNA world" has since been supported.[3] Catalytic DNA molecules, often called deoxyribozymes, DNA enzymes or *DNA*zymes, have thus far not been isolated from any naturally occurring biological system. Because of the absence of the 2'-hydroxyl group in DNA as compared to RNA, it was thought that isolation of catalytically active DNA molecules would pose a much greater challenge than catalytic RNA. However, by the use of an *in vitro* selection technique, Joyce and Breaker demonstrated in 1994 that it was indeed possible to isolate catalytic DNA molecules *in vitro*.[4] Widespread efforts have been since made for the isolation of DNAzymes in laboratories, and these molecules have been engineered to perform various functions. The repertoire of reactions catalyzed by DNAzymes is constantly expanding and includes RNA as well as DNA cleavage and ligation,[4-20] DNA hydrolysis,[21] phosphorylation,[22] capping,[23] cleavage of the phosphoramidite bond,[24] photocleavage of the thymine dimers,[25] deglycosylation,[26] RNA branching and lariat formation,[27-29] and porphyrin metalation[30] in addition to enzymatic activities as peroxidases.[31] Most catalytic NAs isolated thus far exhibit a dependence on metal ion cofactors for carrying out their desired catalytic function and hence have been utilized for metal ion detection.

## 5.2 Detection of Trace Contaminants using Catalytic Nucleic Acids as Sensing Platforms

Detection and quantification of trace contaminants in the environment is an important challenge in the 21st century, since many of these contaminants can pose serious health issues and can cause significant depletion of natural resources. The U.S. Environmental Protection Agency (EPA) has set limits for about 90 contaminants in soil and drinking-water that span a rather wide range of targets, from metal ions, radionuclides, volatile organics, synthetic organics, disinfectants, and their by-products, to viruses, bacteria, and other microbes. Many of these contaminants are present in very low concentrations amidst large amounts of other inorganic and organic species, thereby posing significant challenges in detection due to interferences. Some of the current technologies that are used for trace contaminant detection are based on analytical instrumentation methods that include inductively coupled plasma mass spectrometry (ICP-MS) and atomic absorption spectroscopy (AAS). Both these techniques are very sensitive and selective, and can detect multiple analytes simultaneously. However, these methods are in general rather expensive, require the use of skilled technicians and sample pretreatments, and have long turnaround time, thereby making real-time and on-site detection challenging. To overcome these drawbacks, many sensors have been developed that are both inexpensive and portable. However, there still exists a need for the development of a general sensing platform that can be used for the detection of a wide range and variety of analytes in the environment.

Catalytic NAs possess several advantages as effective and general sensing platforms. Theoretically, they can be engineered toward any desired target cofactor of choice through the combinatorial *in vitro* selection process, which is described in the following section. This method provides a general platform for the isolation and development of a broad range of environmentally relevant targets.[32] Moreover, NAs can be denatured and renatured several times without substantial loss of activity, and are therefore useful for long-term storage in dried and denatured conditions. They can then be regenerated through rehydration in biological buffers prior to use. Modification and labeling for signal transduction is also more predictable with NAs than with proteins or organic molecules, and is therefore useful for rational sensor design. Finally, NAs can be synthesized with different chemical functionalities for relatively low costs. Since DNA in particular is more stable than RNA toward hydrolysis, much easier to chemically synthesize and less expensive, it is a preferred platform for sensor development.[33–36] The focus of this chapter is therefore on catalytic DNA molecules or DNAzymes used for sensing environmentally toxic metal ions.

## 5.3 Isolation of Catalytic Nucleic Acids Using *in vitro* Selection

Catalytic NAs are isolated in the laboratory using *in vitro* selection (Figure 5.1, upper panel). The starting point of this method is the use of large populations ($10^{14}$–$10^{15}$) of chemically synthesized DNA or RNA with random sequences that are iteratively subjected to selection pressure to attain the desired catalytic function.[19–22] After every round of selection, sequences that display the desired catalytic activity are separated from the other DNA/RNA molecules, commonly by various methods that include gel electrophoresis, column-based separations and capillary electrophoresis. Once the desired sequences have been identified, they are amplified using the polymerase chain reaction (PCR) so as

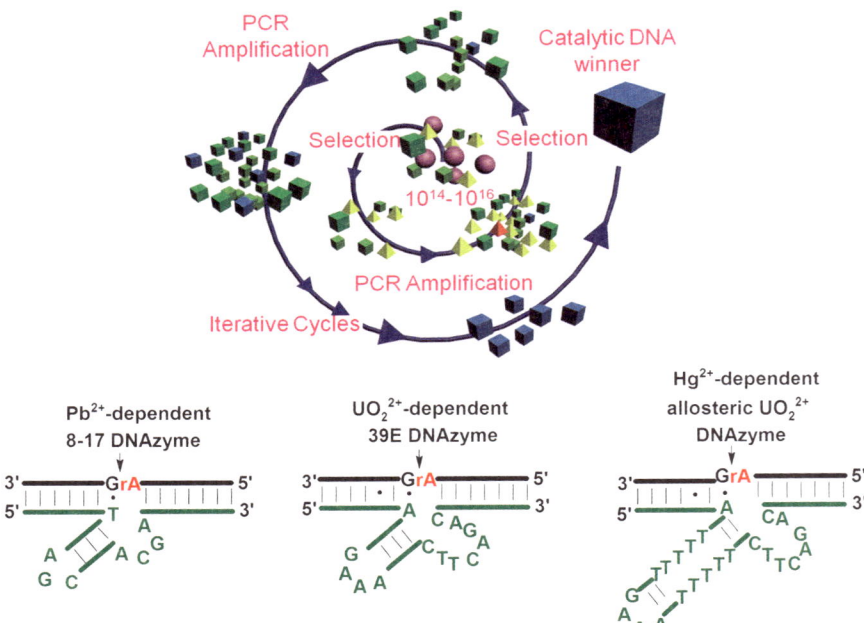

**Figure 5.1** Scheme of the *in vitro* selection method used for isolation of catalytic NAs is shown in the upper panel. The predicted secondary structure of metal ion-dependent RNA-cleaving DNAzymes is shown in the lower panel. The green line represents the enzyme strand, the black line represents the substrate strand, and the RNA cleavage site is shown in red.

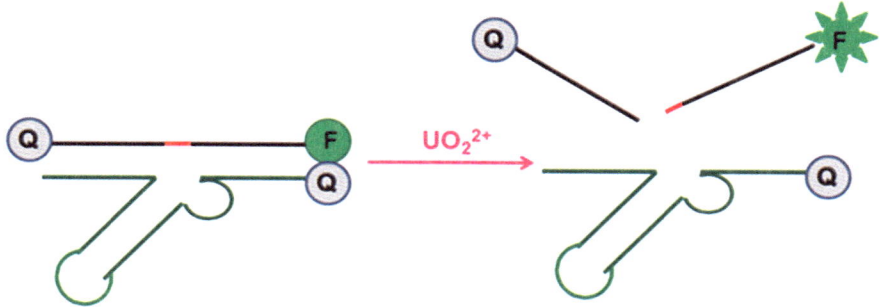

**Figure 5.2** Schematic representation of the 39E DNAzyme-based catalytic beacon sensor for the $UO_2^{2+}$ ion. F, fluorophore; Q quenchers.

to enrich the population of the functional pool. Iterative selection rounds followed by amplification are subsequently carried out and the "winner" sequences exhibiting the highest activity are identified, sequenced and then characterized. Stringency is introduced during the process of *in vitro* selection by lowering either the concentration of the target or the incubation time. Negative selections can also be engineered during the selection procedure to isolate NA sequences that exhibit activity in the presence of competing metal ions. In this case, NA sequences that bind the competing metal ions are discarded and the unbound sequences are retained and utilized for further selection rounds. Selection pressure and stringency can be systematically applied through the *in vitro* selection process to further guide the selection process toward the desired concentration of the metal ion cofactor, thereby increasing the metal ion sensitivity of these catalytic NA molecules and making them more amenable toward metal ion detection for practical applications.

Most of the *in vitro* selected DNAzymes that have been applied for metal ion sensing are RNA-cleaving DNAzymes.[33–36] They include the $Pb^{2+}$-specific 8–17 DNAzyme,[7,11,37,38] the classical $Pb^{2+}$ dependent DNAzyme,[4,39] the $UO_2^{2+}$-specific 39E DNAzyme,[19] an allosteric DNAzyme for $Hg^{2+}$ based on the $UO_2^{2+}$-dependent 39E DNAzyme,[40] and DNAzymes that are dependent on $Co^{2+}$ and $Zn^{2+}$ (Figure 5.2, lower panel).[11,41,42] However, DNA-cleaving DNAzymes specific for $Cu^{2+}$,[6,9] and a $Cu^{2+}$ dependent ligase,[5] have also been isolated.

## 5.4 Conversion of Catalytic Nucleic Acids into Biosensors

The metal ion dependence exhibited by catalytic NAs for activity serves as the basis for their utility as sensing molecules.[43] In the absence of any modifications, these NAs do not possess a strong spectral signal that can be conveniently detected by portable instruments or the naked eye. To transform them into sensors, these molecules therefore have to be combined with signal-producing moieties such as fluorophores, chromophores or electrochemical agents to convert the analyte binding and catalytic activity into detectable signals.

Since the *in vitro* selection strategy involves a predefined secondary structure as the starting point the resulting DNAzymes possess very similar secondary structures (see Figure 5.1), with the only difference being different sequences for different metal ion selectivity. This common secondary structure gives the functional NAs a unique advantage in that, once a general signal transduction method is demonstrated, it can be applied to many other DNAzyme systems for sensing other metal ions. Some of the methods used for the detection of metal ion targets are discussed in the following sections.

## 5.4.1 Fluorescence Sensing

Fluorescence-based detection provides an excellent platform for the development of sensors because of the high sensitivity levels achieved using this method. In addition, there are portable fluorometers available that allow for real-time and on-site detection, thereby adding to the ease of monitoring environmentally relevant contaminants. To utilize these advantages, several functional NAs have been labeled with fluorophores and quenchers and converted into fluorescent sensors.[34,36,44–48]

### 5.4.1.1 Solution-based Fluorescent DNAzyme Sensors

A common strategy utilized in sensing applications of RNA-cleaving DNAzymes is based on the catalytic beacon method.[36,49] An example of this strategy is the 39E DNAzyme that can specifically detect $UO_2^{2+}$.[19] The sensor is assembled by hybridizing the enzyme strand containing a single quencher on the 3′ end and a substrate strand containing an RNA base (that undergoes cleavage), a fluorophore on its 5′ end and a quencher on its 3′ end (Figure 5.2). In the absence of the metal ion, the fluorescence is quenched due to proximity of the fluorophore with the quencher. In the presence of the metal ion, however, cleavage of the RNA base takes place, leading to the release of the substrate arm containing the fluorophore in solution and an overall increase in the fluorescence intensity (∼15 times in this case). The detection limit for $UO_2^{2+}$ based on this method was found to be 45 pM, which was considerably lower than the maximum contamination level (MCL) of 130 nM defined by the U.S. Environmental Protection Agency (EPA), and a dynamic range up to 400 nM. In addition, this sensor also demonstrated more than 1 million-fold selectivity over the other metal ions tested. Finally, when the sensor was used to test contaminated soil samples, the concentrations of $UO_2^{2+}$ obtained were comparable to those obtained using ICP, thereby demonstrating the use of these sensors in practical applications. Previously, a catalytic beacon sensor based on the 8–17 DNAzyme was also shown to be useful for detection of $Pb^{2+}$ in water samples and the detection limit achieved in this case was 10 nM, lower than the U.S. EPA threshold for $Pb^{2+}$ in water which is 72 nM.[50,51] The sensor demonstrated a dynamic range from 10 nM to 4 µM in solution and was also shown to quantify

$Pb^{2+}$ levels in Lake Michigan water samples spiked with $Pb^{2+}$.[51] Through the introduction of mismatches on the enzyme strand of the 17E DNAzyme, it was also shown to detect $Pb^{2+}$ independent of temperature from 4 °C to 30 °C.[52]

The strength of the catalytic beacon sensor method is the generality of the design for application to different DNAzyme systems for the detection of various environmental contaminants. For example, fluorescent sensors for $Cu^{2+}$ and an allosteric sensor for $Hg^{2+}$ have also been demonstrated using this catalytic beacon sensor approach.[40,53] Fluorophore and quencher labels can be positioned at the ends of the substrate and enzyme strand respectively as discussed or can also be incorporated adjacent to the cleavage site directly during the beginning of the *in vitro* selection process, as demonstrated by Li and coworkers.[14,54] These fluorescently labeled DNA molecules can be directly used for sensing applications at the end of the *in vitro* selection process.[34,55]

A novel method of increasing the sensitivity and signal amplification of the fluorescence detection technique has been achieved through a combination of the catalytic beacon and molecular beacon sensor (CAMB).[56] In this study, the substrate strand of the 8–17 DNAzyme was incorporated into the molecular beacon loop, the ends of which were labeled with a fluorophore and a quencher, and the enzyme strand was added to this complex. Addition of $Pb^{2+}$ caused a cleavage of the substrate leading a release of the molecular beacon arms. In the presence of excess molecular beacon substrate, each DNAzyme strand could catalyze multiple cleavage reactions, leading to a significant amplification of the signal. The detection limit of $Pb^{2+}$ achieved using this CAMB system was found to be 600 pM, which was significantly lower than other catalytic beacon methods. The CAMB system was further extended for the detection of adenosine through the incorporation of its respective aptamer sequence into the $Mg^{2+}$-specific 10–23 DNAzyme, to create an allosteric aptazyme system[56] as discussed in section 5.5.

A label-free method based on fluorescence detection has also been recently demonstrated by Lu and coworkers for the detection of $Pb^{2+}$ based on the 8–17 DNAzyme.[57] Label-free sensors generally cost less than labeled methods and cause less interference to the DNAzyme activity in the absence of covalent labeling. In this instance, an abasic site called the dSpacer was introduced on the base-paired binding arms of the enzyme strand, and this site could bind a fluorescent molecule called 2-amino-5,6,7-trimethyl-1,8-naphthyridine (ATMND) in the absence of $Pb^{2+}$ and thereby quench its fluorescence. Upon addition of $Pb^{2+}$, the single RNA base was cleaved, thereby leading to a release of the substrate arm from the enzyme strand as well as the release of the fluorescent molecule. The overall fluorescent intensity was significantly enhanced and the detection limit obtained using this method was found to be 4 nM, which was lower than the labeled sensor, however, the selectivity of the 8–17 DNAzyme was retained. Even though the use of an abasic site is label-free, it still requires a modified nucleotide which is relatively more expensive than its unmodified counterpart. To overcome this limitation, unmodified DNAzymes with a vacant site have been used to demonstrate binding to an extrinsic fluorophore selectively and specifically in the presence of their respective metal ion cofactors.[58]

### 5.4.1.2 Fluorescent DNAzyme Sensors on Solid Supports

Immobilization of fluorescent DNAzyme sensors on solid supports provides an alternate method of sensor design and greater ease of use for real-time detection. Artifacts observed during solution-phase fluorescent studies such as background fluorescence due to excess unhybridized substrate strands can be conveniently eliminated through multiple washing and rinsing steps and the sensors developed can be regenerated and stored over long periods.

The $Pb^{2+}$-specific 8–17 DNAzyme and the $UO_2^{2+}$-specific 39E DNAzyme have been covalently linked to different surfaces including gold,[59] as well as the pores of gold-coated nanocapillary membranes (NCAMs)[60,61], in colloidal assemblies[62] and in hybrids with carbon nanotubes.[63] The immobilized fluorescent DNAzyme sensor on the gold surface yielded a detection limit of 1 nM for $Pb^{2+}$ due to the significantly lowered background, which was accomplished by rinsing away the excess substrate strand. To further increase the surface coverage of the sensor, the DNAzyme sensors were immobilized on the pores of the NCAMs in place of the planar gold.[61] The detection limit achieved in this instance was 17 nM; however, the sensor materials could be regenerated four times with just the addition of new substrate strands and could also be stored at room temperature for about 1 month.

Ye and coworkers have also recently developed a microarray method for the multiplex determination of $Cu^{2+}$ as well as $Pb^{2+}$, based on their respective DNA-cleaving and RNA-cleaving DNAzymes.[64] The substrate strands in this instance were initially immobilized on aldehyde-coated glass slides using Schiff-base chemistry followed by hybridization with the enzyme strand. A Cy-3 fluorophore labeled probe was used to monitor the activity of the sensor in this instance while a Cy-5 labeled probe was used as a reference to normalize the background of the microarray data. In the absence of the metal ion, the fluorescence of the Cy-3 fluorophore was observed since there was no cleavage and the probe remained hybridized to the substrate, however, in the presence of the metal ions (either $Pb^{2+}$ or $Cu^{2+}$) cleavage of the substrate strand took place, resulting in its dissociation into two fragments, thereby leading to the dissociation of the Cy-3 labeled probe strand as well. The sensitivity achieved using this method was approximately 10 nM, the dynamic range was 10 nM–100 μM and considerable selectivity over other metal ions was also observed. This method was also used to test river water samples spiked with two different concentrations of metal ions and showed good recovery in comparison with a cyclic voltammetry (CV) technique.

### 5.4.1.3 Fluorescent DNAzyme based Micro- and Nanofluidic Devices

To extend the applicability of the DNAzyme platforms for sensing applications, they have also been incorporated into the design of micro- and nanofluidic devices. This significantly reduces the usage of the detection reagents and exhibits capabilities for automatic regeneration through computer programming, thereby enabling unattended and long-term monitoring and reducing user risks. The real-time detection of $Pb^{2+}$ based on the fluorescent 8–17 DNAzyme has been successfully demonstrated in microfluidic systems[65] through their immobilization on gold-coated nanocapillary array membranes (NCAM)[66] and on polymethylmethacylate (PMMA) microchannel walls.[67] The sample volumes of the DNAzyme used for $Pb^{2+}$ sensing in these cases was less than 1 nL and a detection limit of 11 nM was achieved with NCAM while the PMMA method yielded a detection limit of 17 nM.[66,67]

### 5.4.2 Colorimetric Sensors

The advantages offered by fluorescent sensors are significant in terms of sensitivity and quantification; however, colorimetric sensors can minimize or even eliminate the need for analytical instrumentation, thereby providing greater ease of use for on-site applications. Since NAs do not exhibit any absorption in the visible region, they are often combined with noble metal nanoparticles such as gold nanoparticles (AuNPs) that exhibit high extinction coefficients ($\sim 2$–3 orders of magnitude higher than organic dyes) and optical properties such as color, which are distance-dependent.[68,69]

DNA-functionalized nanoparticles have therefore been used to demonstrate the sensing of both $Pb^{2+}$ and $UO_2^{2+}$, based on the 8–17 DNAzyme and the 39E DNAzyme respectively.[70–73] Sensing is based on the color change exhibited by the AuNPs, which are blue or purple when aggregated and red when dispersed due to the surface plasmon effect. Moreover, functionalization of AuNPs with DNA can be conveniently obtained or carried out through chemically modified thiol groups. For instance, a colorimetric sensor specific for the detection of $Pb^{2+}$ was carried out based on the functionalization of AuNPs with the 8–17 DNAzyme (Figure 5.3).[70,74] In this case, the substrate strand was extended on either side to facilitate binding to the DNA functionalized on the AuNPs, which was complementary to the former. In the absence of $Pb^{2+}$, a blue color was therefore observed, owing to the aggregation of the AuNPs upon annealing with the DNAzyme. Each of these aggregates contained thousands of AuNPs linked to the DNAzyme. Upon addition of $Pb^{2+}$, cleavage at the RNA base occurred, leading to the dissociation of the substrate strand into two fragments and subsequent disruption of the blue AuNP aggregates into its dispersed state, which changes from blue to red with increased concentrations of $Pb^{2+}$. This color change was significant only in the presence of $Pb^{2+}$ in comparison with the other metal ions tested, and a detection limit of 100 nM was achieved. Colorimetric sensors based on this design were, in fact, used to detect $Pb^{2+}$ extracted from paint samples.[69] The dynamic range of this sensor was tuned through the introduction of an inactive 8–17 DNAzyme (containing a G-C base pair in place of a G·T wobble pair) which was still capable of assembling the AuNPs with efficiency almost similar to the original DNAzyme.[71] The response of this sensor system then moved to a higher order of magnitude, simply through the introduction of about 95% inactive DNAzyme. Different ratios of

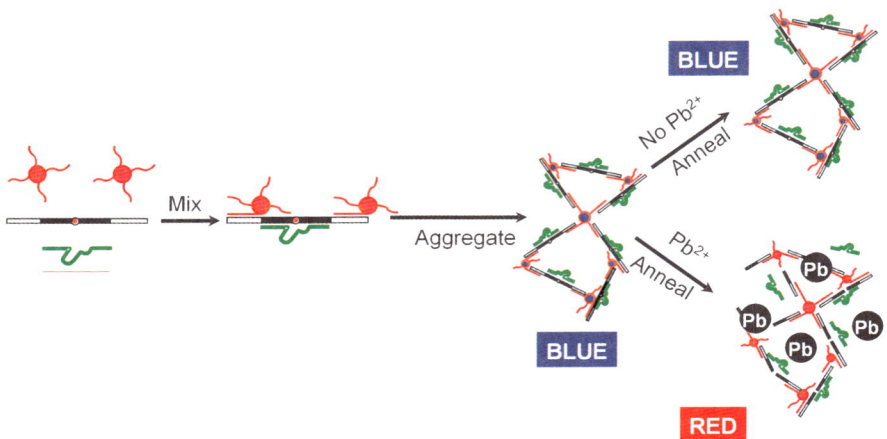

**Figure 5.3** Schematic representation of the labeled 8–17 DNAzyme-based colorimetric sensor for $Pb^{2+}$ detection. In the absence of $Pb^{2+}$, the DNA-functionalized AuNPs are assembled on the substrate to form blue aggregates. When $Pb^{2+}$ is present, the substrate is cleaved and the aggregate is disassembled to yield red colored dispersed AuNPs. Adapted with permission from J. Liu and Y. Lu, *J. Am. Chem. Soc.*, 2003, **125**, 6642. Copyright 2010 American Chemical Society

the active and the inactive DNAzyme could then be used, particularly for instances when higher detection ranges were needed, thereby tuning the overall response of the sensor (Figure 5.3).

To improve the reaction rate, the diameter of the AuNPs was increased from 13 nm to 42 nm leading to a distinct color change in about 5 min, and in addition, small invasive DNAs were also designed and introduced to improve sensing.[71,74,75] Colorimetric sensors for $UO_2^{2+}$ have also been developed based on this design and the color change from blue to red has been observed in the presence of this radionuclide, while such a change has not been observed with the other metal ions tested. The detection limit achieved in this case was 50 nM with a detection range from 50 to 2 µM.[76]

Label-free colorimetric sensors for both $Pb^{2+}$ and $UO_2^{2+}$ have also been designed.[76,77] The basis of sensing in these cases is the difference in adsorption of single-stranded (ss) DNA versus that of double-stranded (ds) DNA on bare AuNP surfaces, leading to a change in the stability of the nanoparticles in the presence of NaCl. In the absence of the metal ion the dsDNA cannot adsorb onto the surface of the AuNPs, and at a certain concentration of NaCl they aggregate together, leading to a blue solution. In the presence of the metal ion, cleavage at the RNA base takes place leading to the release of the ssDNA fragment that is adsorbed onto the surface of the AuNPs, thereby preventing their aggregation. The color of the solution then is red, owing to the dispersed particles. The detection limit achieved in this case was found to be 3 nM for $Pb^{2+}$,[77] and 1 nM for $UO_2^{2+}$.[76]

Colorimetric sensing of $Pb^{2+}$ and $UO_2^{2+}$ based on both the labeled and label-free methods has been demonstrated and a detailed comparison of the two designs has been previously discussed.[76,77] Briefly, the labeled sensor required longer preparation times and showed a higher detection limit for $UO_2^{2+}$; however, the ease of use of operation was significantly higher than for the label-free system, which was relatively easy to prepare and use and also showed higher sensitivity toward $UO_2^{2+}$. Nonetheless, the label-free method was more vulnerable to external variables as well as the ionic strength. A highly sensitive ligation-based colorimetric sensor for $Cu^{2+}$ using a ligation DNAzyme has also been demonstrated, and this system has been shown to be sensitive enough to detect $Cu^{2+}$ in drinking water without the need for analytical instrumentation.[78] Since the DNAzyme used in this case was based on a ligation reaction rather than a cleavage reaction, it had an intrinsically lowered background, thereby leading to higher sensitivity.

### 5.4.3 Dipstick Tests based on AuNP-DNAzyme Conjugates

While the use of colorimetric sensors have shown to minimize or even eliminate the need for analytical instrumentation and thereby facilitate on-site detection, laboratory operations such as the precise transfer and mixing of microliter quantities of solutions are still essential. Moreover, the distinction of color upon increasing the metal ion concentration, going from the blue and purple aggregates to the red dispersed solution, is incremental and can be rather difficult to distinguish by the naked eye. Storage of the colorimetric sensors over long periods of time is also difficult because AuNPs aggregate in solution. Therefore, to further facilitate the ease of use of these sensor materials for nontechnical users, lateral flow devices based on the 8–17 DNAzyme for $Pb^{2+}$ detection have been recently developed,[79] following successful demonstration of dipstick tests for adenosine and cocaine.[80]

The 8–17 DNAzyme used for $Pb^{2+}$ detection (Figure 5.4a) was modified for optimal activity on the dipstick (Figure 5.4b).[79] It was biotinylated on the 3' end and was extended and thiolated on the 5' end of the substrate arm to facilitate functionalization onto the surface of a AuNP. The lateral flow device was composed of four regions, consisting of four overlapping pads, namely an absorption pad, a membrane pad, a conjugation pad and a wicking pad (Figure 5.4c). The membrane pad consisted of two zones, a control zone on which streptavidin was spotted and a test zone on which capture DNA, complementary to the cleaved substrate DNA fragment was spotted. The modified 8–17 DNAzyme construct was then spotted onto the conjugation pad and allowed to dry for about 8 h. Upon dipping the wicking pad into a flow buffer solution, the liquid moved up, rehydrating the sensor functionalized on the AuNPs. In the absence of $Pb^{2+}$, no cleavage of the substrate was observed and the complex migrated till the control zone wherein the biotinylated DNA was captured by the streptavidin, which was indicated by the appearance of a single red line (Figure 5.4d). When the same test was carried out in the presence of

Pb$^{2+}$ in the flow buffer, cleavage of the substrate took place, leading to the appearance of the first red line at the capture zone corresponding to the capture of the biotinylated DNA, and a second red line, corresponding to the capture of the cleaved DNA fragment by its complementary DNA or the capture DNA at the test zone (Figure 5.4e). The detection limit achieved by this method was approximately 5 µM and the reaction time was about 10 min. The detection limit achieved by this method was lowered further to 0.5 µM by performing the cleavage reaction in solution in the presence of Pb$^{2+}$ followed by dipping in the flow buffer. The practical applicability of this test was also determined by

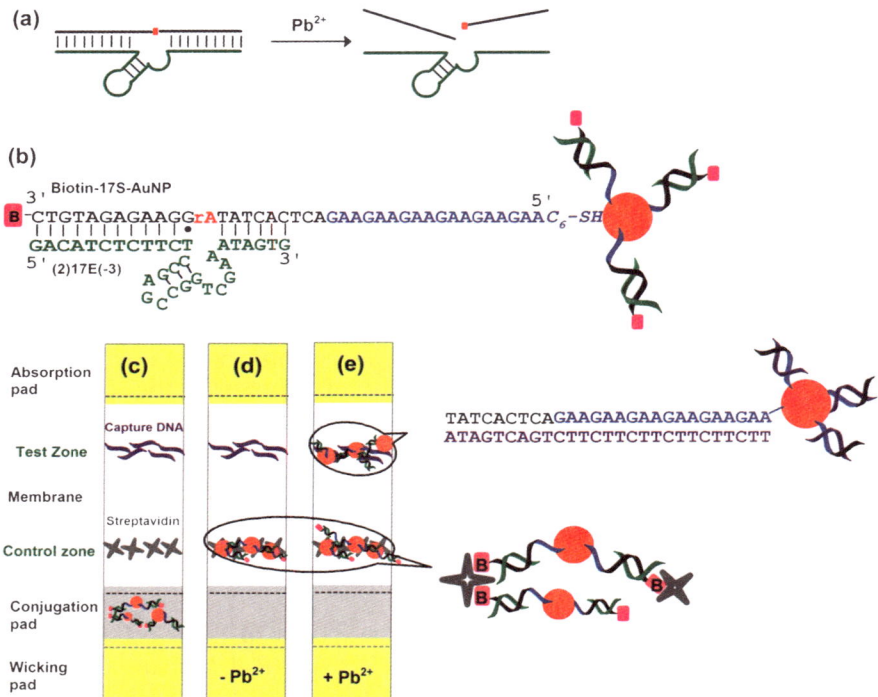

**Figure 5.4** Scheme showing the lateral flow device for detection of Pb$^{2+}$, adapted from Mazumdar et al.[79] (a) 8–17 DNAzyme reaction: In the presence of Pb$^{2+}$, the 17E enzyme (green line) catalyzes the cleavage of the substrate, 17S (black line) at the sigle ribo-linkage (shown in red) to give two product strands. (b) Modified 8–17 construct used for the dipstick tests. The substrate is functionalized on the gold nanoparticles and contains a biotin moiety, indicated by the pink square on its 3′ end. (c) Assembled lateral flow device: Streptavidin and capture DNA are applied to the control and test zone of the membrane respectively and the Enz-SubAuNP is placed on the conjugation pad. (d) Negative test producing a single red line at the control zone due to accumulation of nanoparticles. (e) Positive test producing two red lines at the control and test zones.

detecting the amount of $Pb^{2+}$ in paints which were previously spiked with different amounts of $Pb^{2+}$. EDTA, a good chelator of $Pb^{2+}$ was also added to tune the dynamic range of the sensor leading to an approximately fivefold increase of the detection limit and elimination of any nonspecific background cleavage. (Figure 5.4).

### 5.4.4 Electrochemical Sensors

Electrochemical methods for the detection of metal ions based on catalytic NAs have also been developed as alternative approaches to the optical detection methods. These sensors are particularly attractive in terms of their practical applications since they do not suffer from background interference or false signals and can also be easily converted into miniature versions in terms of their size and power. These devices are also reliable and provide consistent results with high reproducibility, sensitivity and fast response times. Plaxco and coworkers have demonstrated the electrochemical method of sensing $Pb^{2+}$ by immobilizing the enzyme strand of the 8–17 DNAzyme onto a gold electrode through its 5′ end and using a methylene blue (MB) reporter group on the 3′ end, which is hybridized to a complementary substrate strand containing the single RNA cleavage site.[81] The MB reporter is designed to be at a distance from the electrode surface due to the rigidity of the enzyme–substrate complex in the absence of the metal ion, therefore leading to a low electron transfer rate. In the presence of $Pb^{2+}$, however, the cleavage of the substrate followed by its dissociation into two fragments took place leading to a greater flexibility of the enzyme containing the MB complex, thereby promoting electron transfer. High selectivity for $Pb^{2+}$ was obtained and a detection limit of 300 nM was achieved. The ability of this sensor to detect $Pb^{2+}$ in soil samples spiked with the metal ion was also demonstrated. Shao and coworkers have combined the electrochemical detection of $Pb^{2+}$ based on the 8–17 DNAzyme using a $Ru(NH_3)_6^{2+}$ reporter molecule, with gold bio-barcodes, which are short oligonucleotides functionalized on a 13 nm AuNP that are complementary to the extended substrate strand.[82] The use of these bio-barcodes significantly amplified the signal leading to the enhancement of the sensitivity of the method, and a detection limit of 1 nM was achieved. An electrochemiluminescent sensor for $Pb^{2+}$ based on the 8–17 DNAzyme and $Ru(bpy)_3^{2+}$ has also been recently developed by Chen and coworkers and the lowest detection limit based on the 8–17 DNAzyme of 0.1 nM has been reported.[83]

## 5.5 Expanding the Scope of the Sensing Targets of Catalytic Nucleic Acids by Employing Aptazymes

The repertoire of reactions catalyzed by NA enzymes has been extended to recognition of molecular targets through their combination with aptamers, and these combined functional NAs have been referred to as allosteric RNA/

DNAzymes or aptazymes.[84] The aptamer module serves as an allosteric regulator of the catalytic activity of the DNAzyme module, analogous to the regulation of protein enzymes. While allosteric DNAzymes have thus far been obtained through the deliberate combination of known aptamers and catalytic NAs, allosteric ribozymes have been isolated through *in vitro* selection as well.[84] An aptazyme based on the 8–17 DNAzyme and the adenosine aptamer, wherein the aptamer strand was introduced into one of the substrate binding arms of the DNAzyme, was utilized for the colorimetric detection of adenosine through the functionalization of AuNPs with a modified 8–17 aptazyme.[85] In the absence of the adenosine, there was no substrate cleavage in the presence of $Pb^{2+}$ and a blue color owing to the aggregation of AuNPs was observed. However, in the presence of adenosine, the catalytic conformation of the 8–17 DNAzyme was restored, leading to a cleavage of the substrate in the presence of $Pb^{2+}$ and a color change from blue to red. This color change was not observed in the presence of other nucleotides studied. The use of fluorescent signaling aptazymes has also been demonstrated by Li and coworkers. The catalytic cleavage of the DNAzyme module is controlled by competition between an external antisense oligonucleotide and the molecular aptamer target, namely ATP.[86] Interestingly, aptamers for ATP have also been shown to regulate the inhibition of catalytic activity of aptazymes.[87] The CAMB system has also been used to demonstrate the detection of adenosine through the fusion of the adenosine aptamer with the 10–23 DNAzyme which is added to the molecular beacon containing the corresponding substrate.[56] The absence of adenosine leads to an inhibition of the core structure of the DNAzyme essential for catalytic activity. In the presence of adenosine, however, the core is structured, leading to an activation of the DNAzyme and catalysis in the presence of $Mg^{2+}$ with a detection limit of 500 nM for adenosine, which is the lowest reported for a fluorescence-based adenosine sensing system. Signal amplification can also be achieved through the introduction of excess substrate to achieve multiple turnovers.[56] Aptazyme-based sensing using the 8–17 DNAzyme in combination with the adenosine aptamer has also been reported using an electrochemical method.[88]

A recent study has also demonstrated the use of a DNAzyme in place of an aptamer to create an allosteric dual-DNAzyme system for the detection of $Cu^{2+}$ in drinking-water.[89] In this instance, a DNA-cleaving DNAzyme containing the substrate strand was fused with an HRP-mimicking DNAzyme. In the presence of $Cu^{2+}$ cleavage at the substrate strand was achieved, which in turn activated the HRP-mimicking DNAzyme to elicit a colorimetric signal with hemin. Detection limits in the ppb range for $Cu^{2+}$ were achieved using this method.[89] Colorimetric sensing has also been achieved through the coupling of an allosteric RNA-cleaving DNAzyme with the rolling circle amplification (RCA) technique to generate long single strands of DNA which can further hybridize with complementary PNA strands in the presence of a chromophoric duplex-DNA binding dye which changes color from blue to purple upon binding.[90]

## 5.6 Summary and Future Perspectives

Since their discovery, catalytic NAs have attracted considerable attention because of their capability of catalyzing a wide variety of reactions. Moreover, since the catalytic function of most known DNAzymes are metal-ion dependent, they have also gained attention as practical alternatives for the detection of metal ions—in particular, environmentally relevant metal ions. DNAzymes for any environmentally relevant metal ion can theoretically be obtained using the *in vitro* selection method. Sensitivity and selectivity can also be fine-tuned through the variation of selection conditions to detect even minute concentrations of these metal ions in the environment. Upon combination of these DNAzymes with suitable signal transduction methods such as fluorescence, colorimetric methods based on AuNPs and electrochemical methods described here, metal ion-specific sensors can be developed as practical alternatives for real-time and on-site detection, almost all of which have detection limits much lower than the MCL defined by the U.S. EPA. In fact, sensors for the detection of heavy metal ions such as $Pb^{2+}$ and $UO_2^{2+}$ are now commercially available (http://www.andalyze.com/) less than 10 years since the first report of the DNAzyme-based metal ion sensor.[50]

Despite the advances in the development of catalytic NA sensors, a major challenge lies in expanding the current range of metal ions to other environmentally relevant ionic targets that are negatively charged, such as the perchlorate and nitrate anions. In addition, there is significant room for improvement of the *in vitro* selection method. Efforts toward automating the selection process have been reported to improve its speed and efficiency.[91,92] Sensor array development for simultaneous detection of multiple analytes is also an important step in furthering the practical applicability of these catalytic DNA molecules.

## Acknowledgments

The authors would like to thank Dr. Debapriya Mazumdar for helpful suggestions, and the U.S. National Institutes of Health (ES16865), the Department of Energy (DE-FG02–08ER64568), and the National Science Foundation (CTS-0120978, DMI-0328162 and DMR-0117792) for financial support.

## References

1. K. Kruger, P. J. Grabowski, A. J. Zaug, J. Sands, D. E. Gottschling and T. R. Cech, *Cell*, 1982, **31**, 147.
2. C. Guerrier-Takada, K. Gardiner, T. Marsh, N. Pace and S. Altman, *Cell*, 1983, **35**, 849.
3. U. F. Muller, *Cell Mol. Life Sci.*, 2006, **63**, 1278.
4. R. R. Breaker and G. F. Joyce, *Chem. Biol.*, 1994, **1**, 223.
5. B. Cuenoud and J. W. Szostak, *Nature*, 1995, **375**, 611.
6. N. Carmi, L. A. Shultz and R. R. Breaker, *Chem. Biol.*, 1996, **3**, 1039.

7. S. W. Santoro and G. F. Joyce, *Proc. Natl. Acad. Sci. U. S. A.*, 1997, **94**, 4262.
8. C. R. Geyer and D. Sen, *Chem. Biol.*, 1997, **4**, 579.
9. N. Carmi, H. R. Balkhi and R. R. Breaker, *Proc. Natl. Acad. Sci. U. S. A.*, 1998, **95**, 2233.
10. S. W. Santoro, G. F. Joyce, K. Sakthivel, S. Gramatikova and C. F. Barbas III, *J. Am. Chem. Soc.*, 2000, **122**, 2433.
11. J. Li, W. Zheng, A. H. Kwon and Y. Lu, *Nucleic Acids Res.*, 2000, **28**, 481.
12. D. M. Perrin, T. Garestier and C. Helene, *J. Am. Chem. Soc.*, 2001, **123**, 1556.
13. A. Flynn-Charlebois, Y. Wang, T. K. Prior, I. Rashid, K. A. Hoadley, R. L. Coppins, A. C. Wolf and S. K. Silverman, *J. Am. Chem. Soc.*, 2003, **125**, 2444.
14. S. H. J. Mei, Z. Liu, J. D. Brennan and Y. Li, *J. Am. Chem. Soc.*, 2003, **125**, 412.
15. A. Sreedhara, Y. Li and R. R. Breaker, *J. Am. Chem. Soc.*, 2004, **126**, 3454.
16. A. V. Sidorov, J. A. Grasby and D. M. Williams, *Nucleic Acids Res.*, 2004, **32**, 1591.
17. K. A. Hoadley, W. E. Purtha, A. C. Wolf, A. Flynn-Charlebois and S. K. Silverman, *Biochemistry*, 2005, **44**, 9217.
18. W. E. Purtha, R. L. Coppins, M. K. Smalley and S. K. Silverman, *J. Am. Chem. Soc.*, 2005, **127**, 13124.
19. J. Liu, A. K. Brown, X. Meng, D. M. Cropek, J. D. Istok, D. B. Watson and Y. Lu, *Proc. Natl. Acad. Sci. U. S. A.*, 2007, **104**, 2056.
20. K. Schlosser, J. Gu, J. C. Lam and Y. Li, *Nucleic Acids Res.*, 2008, **36**, 4768.
21. M. Chandra, A. Sachdeva and S. K. Silverman, *Nat. Chem. Biol.*, 2009, **5**, 718.
22. W. Wang, L. P. Billen and Y. Li, *Chem. Biol.*, 2002, **9**, 507.
23. Y. Li, Y. Liu and R. R. Breaker, *Biochemistry*, 2000, **39**, 3106.
24. J. Burmeister, G. von Kiedrowski and A. D. Ellington, *Angew. Chem. Int. Ed.*, 1997, **36**, 1321.
25. D. J. F. Chinnapen and D. Sen, *Proc. Natl. Acad. Sci. U. S. A.*, 2004, **101**, 65.
26. T. L. Sheppard, P. Ordoukhanian and G. F. Joyce, *Proc. Natl. Acad. Sci. U. S. A.*, 2000, **97**, 7802.
27. Y. Wang and S. K. Silverman, *J. Am. Chem. Soc.*, 2003, **125**, 6880.
28. Y. Wang and S. K. Silverman, *Biochemistry*, 2005, **44**, 3017.
29. R. L. Coppins and S. K. Silverman, *Nat. Struct. Mol. Biol.*, 2004, **11**, 270.
30. Y. Li and D. Sen, *Nat. Struct. Biol.*, 1996, **3**, 743.
31. P. Travascio, A. J. Bennet, D. Y. Wang and D. Sen, *Chem. Biol.*, 1999, **6**, 779.
32. D. Mazumdar, J. Liu and Y. Lu, in *Nanotechnology Applications for Clean Water*, ed. N. Savage, M. Diallo, J. Duncan, A. Street and R. Sustich, William Andrews, Norwich, NY, 2009, p. 427.
33. J. C. Achenbach, W. Chiuman, R. P. Cruz and Y. Li, *Curr. Pharm. Biotechnol.*, 2004, **5**, 321.
34. W. Chiuman and Y. Li, in *Functional Nucleic Acids for Analytical Applications*, ed. Y. Li and Y. Lu, Springer, New York, NY, 2009, p. 131.

35. S. K. Silverman, *Nucleic Acids Res.*, 2005, **33**, 6151.
36. J. Liu, Z. Cao and Y. Lu, *Chem. Rev.*, 2009, **109**, 1948.
37. A. K. Brown, J. Li, C. M. B. Pavot and Y. Lu, *Biochemistry*, 2003, **42**, 7152.
38. K. Schlosser and Y. Li, *ChemBioChem*, 2010, **11**, 866.
39. T. Lan, K. Furuya and Y. Lu, *Chem. Commun.*, 2010, **46**, 3896.
40. J. Liu and Y. Lu, *Angew. Chem. Int. Ed.*, 2007, **46**, 7587.
41. P. J. Bruesehoff, J. Li, A. J. Augustine and Y. Lu, *Comb. Chem. High Throughput Screening*, 2002, **5**, 327.
42. K. E. Nelson, P. J. Bruesehoff and Y. Lu, *J. Mol. Evol.*, 2005, **61**, 216.
43. Z. Cao and Y. Lu, in *Metal Complex–DNA Interactions*, ed. N. Hadjiliadis and E. Sletten, Wiley-Blackwell, Oxford, UK, 2009, p. 395.
44. H. Wang, Y. Kim, H. Liu, Z. Zhu, S. Bamrungsap and W. Tan, *J. Am. Chem. Soc.*, 2009, **131**, 8221.
45. N. K. Navani and Y. Li, *Curr. Opin. Chem. Biol.*, 2006, **10**, 272.
46. N. Rupcich, W. Chiuman, R. Nutiu, S. Mei, K. K. Flora, Y. Li and J. D. Brennan, *J. Am. Chem. Soc.*, 2006, **128**, 780.
47. W. Chiuman and Y. Li, *Nucleic Acids Res.*, 2007, **35**, 401.
48. J. Liu and Y. Lu, in *Functional Nucleic Acids for Sensing and Other Analytical Applications*, ed. Y. Li and Y. Lu, Springer, New York, NY, 2009, p. 155.
49. J. Liu and Y. Lu, in *Methods in Molecular Biology*, ed. V. V. Didenko, Humana Press, Totowa, NJ, 2006, p. 275.
50. J. Li and Y. Lu, *J. Am. Chem. Soc.*, 2000, **122**, 10466.
51. J. Liu and Y. Lu, *Anal. Chem.*, 2003, **75**, 6666.
52. N. Nagraj, J. Liu, S. Sterling, J. Wu and Y. Lu, *Chem. Commun.*, 2009, 4103.
53. J. Liu and Y. Lu, *J. Am. Chem. Soc.*, 2007, **129**, 9838.
54. Z. Liu, S. H. J. Mei, J. D. Brennan and Y. Li, *J. Am. Chem. Soc.*, 2003, **125**, 7539.
55. Y. Shen, G. Mackey, N. Rupcich, D. Gloster, W. Chiuman, Y. Li and J. D. Brennan, *Anal. Chem.*, 2007, **79**, 3494.
56. X.-B. Zhang, Z. Wang, H. Xing, Y. Xiang and Y. Lu, *Anal. Chem.*, 2010, **82**, 5005.
57. Y. Xiang, A. Tong and Y. Lu, *J. Am. Chem. Soc.*, 2009, **131**, 15352.
58. Y. Xiang, Z. Wang, H. Xing, N. Y. Wong and Y. Lu, *Anal. Chem.*, 2010, **82**, 4122.
59. C. B. Swearingen, D. P. Wernette, D. M. Cropek, Y. Lu, J. V. Sweedler and P. W. Bohn, *Anal. Chem.*, 2005, **77**, 442.
60. D. P. Wernette, C. B. Swearingen, D. M. Cropek, Y. Lu, J. V. Sweedler and P. W. Bohn, *Analyst*, 2006, **131**, 41.
61. D. P. Wernette, C. Mead, P. W. Bohn and Y. Lu, *Langmuir*, 2007, **23**, 9513.
62. M. H. Shyr, D. P. Wernette, P. Wiltzius, Y. Lu and P. V. Braun, *J. Am. Chem. Soc.*, 2008, **130**, 8234.
63. T.-J. Yim, J. Liu, Y. Lu, R. S. Kane and J. S. Dordick, *J. Am. Chem. Soc.*, 2005, **127**, 12200.
64. P. Zuo, B.-C. Yin and B.-C. Ye, *Biosens. Bioelectron.*, 2009, **25**, 935.

65. K. A. Shaikh, K. S. Ryu, E. D. Goluch, J. Nam, J. Liu, C. S. Thaxton, T. N. Chiesl, A. E. Barron and Y. Lu, *Proc. Natl. Acad. Sci. U.S.A.*, 2005, **102**, 9745.
66. I.-H. Chang, J. J. Tulock, J. Liu, W.-S. Kim, D. M. Cannon Jr, Y. Lu, P. W. Bohn, J. V. Sweedler and D. M. Cropek, *Environ. Sci. Technol.*, 2005, **39**, 3756.
67. T. S. Dalavoy, D. P. Wernette, M. Gong, J. V. Sweedler, Y. Lu, B. R. Flachsbart, M. A. Shannon, P. W. Bohn and D. M. Cropek, *Lab Chip*, 2008, **8**, 786.
68. Y. Lu and J. Liu, *Curr. Opin. Biotechnol.*, 2006, **17**, 580.
69. W. Zhao, M. A. Brook and Y. Li, *ChemBioChem*, 2008, **9**, 2363.
70. J. Liu and Y. Lu, *J. Am. Chem. Soc.*, 2003, **125**, 6642.
71. J. Liu and Y. Lu, *J. Am. Chem. Soc.*, 2005, **127**, 12677.
72. J. Liu and Y. Lu, *Org. Biomol. Chem.*, 2006, **4**, 3435.
73. W. Zhao, J. C. Lam, W. Chiuman, M. A. Brook and Y. Li, *Small*, 2008, **4**, 810.
74. J. Liu and Y. Lu, *J. Am. Chem. Soc.*, 2004, **126**, 12298.
75. J. Liu and Y. Lu, *Chem. Mater.*, 2004, **16**, 3231.
76. J. H. Lee, Z. Wang, J. Liu and Y. Lu, *J. Am. Chem. Soc.*, 2008, **130**, 14217.
77. Z. Wang, J. H. Lee and Y. Lu, *Adv. Mat.*, 2008, **20**, 3263.
78. J. Liu and Y. Lu, *Chem. Commun.*, 2007, 4872.
79. D. Mazumdar, J. Liu, G. Lu, J. Zhou and Y. Lu, *Chem. Commun.*, 2010, **46**, 1416.
80. J. Liu, D. Mazumdar and Y. Lu, *Angew. Chem. Int. Ed.*, 2006, **45**, 7955.
81. Y. Xiao, A. A. Rowe and K. W. Plaxco, *J. Am. Chem. Soc.*, 2007, **129**, 262.
82. L. Shen, Z. Chen, Y. Li, S. He, S. Xie, X. Xu, Z. Liang, X. Meng, Q. Li, Z. Zhu, M. Li, X. C. Le and Y. Shao, *Anal. Chem.*, 2008, **80**, 6323.
83. X. Zhu, Z. Lin, L. Chen, B. Qiu and G. Chen, *Chem. Commun.*, 2009, 6050.
84. R. R. Breaker, *Curr. Opin. Biotechnol.*, 2002, **13**, 31.
85. J. Liu and Y. Lu, *Anal. Chem.*, 2004, **76**, 1627.
86. J. C. Achenbach, R. Nutiu and Y. Li, *Anal. Chim. Acta*, 2005, **534**, 41.
87. Y. Shen, W. Chiuman, J. D. Brennan and Y. Li, *ChemBioChem*, 2006, **7**, 1343.
88. C. Sun, X. Liu, K. Feng, J. Jiang, G. Shen and R. Yu, *Anal. Chim. Acta*, 2010, **669**, 87.
89. B. C. Yin, B. C. Ye, W. Tan, H. Wang and C. C. Xie, *J. Am. Chem. Soc.*, 2009, **131**, 14624.
90. M. M. Ali and Y. Li, *Angew. Chem. Int. Ed.*, 2009, **48**, 3512.
91. L. J. Sooter, T. Riedel, E. A. Davidson, M. Levy, J. C. Cox and A. D. Ellington, *Biol. Chem.*, 2001, **382**, 1327.
92. X. Lou, J. Qian, Y. Xiao, L. Viel, A. E. Gerdon, E. T. Lagally, P. Atzberger, T. M. Tarasow, A. J. Heeger and H. T. Soh, *Proc. Natl. Acad. Sci. U. S. A.*, 2009, **106**, 2989.

CHAPTER 6
# Nucleic Acid-based Biosensors for the Detection of DNA Damage

KIM R. ROGERS[1] AND RONALD K. GARY[2]

[1] Human Exposure and Atmospheric Sciences Division, U.S. Environmental Protection Agency, Las Vegas, NV 89119, USA; [2] Department of Chemistry, University of Nevada Las Vegas, Las Vegas, NV 89154, USA

## 6.1 Introduction

Monitoring the environment for the presence of toxic compounds is not only required for clean-up of previously contaminated land and water, but is important for preventing future contamination and managing risks to human health and the environment. Identification and measurement of specific compounds in the environment that are known to have toxic effects has historically been the strategy used to accomplish this task. The measurement of compounds in environmental media currently involves several steps such as sampling, transport to the laboratory, extraction, sample clean-up and analysis by instrumental techniques. Each of these steps requires time, effort, and expense. This is not a trivial task considering that there are about 80 000 chemicals currently in use and more than 1000 are added to the market every year.[1] In addition to the parent compounds that may be released, there are also a vast number of compounds resulting from the decomposition of original pollutants. Given that only a small fraction of the vast number of compounds that contaminate the environment from anthropogenic processes are toxicologically well described, determining and managing potential risks by identifying and monitoring specific compounds is a considerable undertaking.

One alternative approach to rapidly identify the presence of potentially harmful substances in environmental matrices is through the use of toxicity screening assays. Although this concept is not new (*e.g.*, the canary in the coal mine), advances in biology, biochemistry, and sensor technology have resulted in a wide range of bioanalytical toxicity screening assays. With respect to the diversity in mechanisms of chemical toxicity, there have also been a number of approaches developed to screen for toxicity using a range of indicator organisms and biochemically based formats. Because any particular organism may not be representative of the physiology and biochemistry of other organisms, even within a similar ecological niche, batteries of screening tests have been adopted.[2] Partly for practical and economic reasons, the trend in recent decades has been to move from more complex systems such as mammals and fish to simpler systems that retain sensitivity such as embryonic organisms, microinvertebrates, algae, and microorganisms. Advances in microelectronics and microfluidics have also resulted in the trend toward the use of multiple tests being conducted in a microarray format, to potentially screen for toxic effects using culture cells, enzymes, and nucleic acids.[3–6]

One mechanism for toxicity that has been linked to carcinogenicity and potential problems for future generations is *genotoxicity*. Similarly to screening for cytotoxicity, the measurement of genotoxicity has also relied on a battery of assays that are able to identify a majority of rodent tumorigenic compounds. These assays currently include a bacterial reversion assay (Ames), an *in vitro* mouse lymphoma assay, and assays to measure chromosomal aberrations in mammalian culture cells.[5]

In addition to assays that measure mutations or chromosomal aberrations, there are also a range of bioassays that measure chemical damage to DNA. Oxidative damage resulting in the formation of 8-oxo-7,8-dihydroguanine has been detected using gas chromatography-mass spectrometry (GC-MS) or high performance liquid chromatography (HPLC) with electrochemical detection.[7] General assays for DNA damage also include the alkaline elution assay, alkaline unwinding assay, and comet assay.[4] These general assays depend on the chemically induced partial unwinding of DNA to distinguish damaged from control DNA. This structural amplification allows these assays to be particularly sensitive to alkylation, adduct formation, cross-linking, and oxidative damage.[4]

Although many of the mutagenesis assays and assays for DNA damage are referred to as short-term screening methods, they often take hours to days to complete, are expensive, and require sophisticated technical expertise to run. As a result, they are not typically well suited to adaptation to *field screening applications*. By contrast, rapid, inexpensive, and simple biosensor screening assays for DNA damage may not be as selective or sensitive to clastogens or mutagens; however, these assays appear to be potentially well suited for screening large numbers of samples or continuous field monitoring applications. Consequently, assays that fit the biosensor concept may complement other more complex and time-consuming screening assays in forming a battery of screening tests for environmental applications.[8]

The generally accepted *definition of a biosensor* is a self-contained integrated device consisting of a biological recognition element (*e.g.*, enzyme, antibody, aptamer, nucleic acid, microorganism, *etc.*) which is interfaced to a physical transducer that converts the recognition event into a measureable signal.[3,6,9] Biosensors are typically differentiated from bioanalytical systems in that the latter lack a direct interface of the recognition element to the detector, and they often require multiple reagents and multiple steps. Nevertheless, because a number of bioanalytical assay systems show potential for development as biosensors and because some of these systems are approaching the capability for self-contained and autonomous function in environmental settings, these techniques will also be included in this chapter. For example, "lab-on-a-chip" techniques may create some of the same capabilities for these devices as have been envisioned for biosensors (*e.g.*, inexpensive per assay, durable, reliable, and capable of remote operation). In any case, the promise of rapid, simple, inexpensive, sensitive, continuous, and remotely operated analytical devices continues to push the envelope for development of biosensors for potential environmental applications.[6]

Biosensor-based genotoxicity assays and rapid screening assays for potential environmental applications have primarily focused on two types of systems. These systems involve the measurement of chemically induced damage to *surrogate DNA* and the responses of genetically engineered microorganism reporters that express proteins or processes that can be measured when these organisms are challenged by DNA-damaging chemicals. Biosensors and rapid screening assays that respond to a wide range of compounds that are known to cause DNA damage by means of a number of mechanisms have been reported. Nevertheless, the adaptation of biosensors for the measurement of genotoxins in the environment faces a number of practical as well as conceptual challenges. The first assumption that must be made for most biosensors is that chemically induced damage to surrogate DNA will be an acceptable screening model relevant to the wide range of organisms (including mammalian systems) that may potentially be exposed to polluted environmental media. These types of screening assays are typically afforded a relatively large latitude with respect to exposure and dose metrics. More specifically, the value of these methods would be primarily to alert or warn of the presence of potentially genotoxic substances in a particular sample. This is not an inconsequential matter, given that these compounds might be located in relatively few high-concentration "hot spots" spread over a large area involving potentially thousands of samples. Screening techniques can also be of significant value for monitoring for discrete releases of genotoxins in a continuous stream.

Another assumption that must be made with respect to biosensors that measure DNA damage is that the compounds causing the damage to surrogate DNA would find their way (in sufficient concentration) to the most sensitive nucleotide sequences in the cell without metabolic modification or detoxification. In order to better approximate enzymatically catalyzed activation of specific genotoxins, rapid bioanalytical assays and biosensor techniques have incorporated *pretreatments* with liver enzymes in the form of microsomes (S-9 fraction) or isolated cytochrome P450 enzymes.[10] For example, arrays of DNA and enzyme-containing thin film spots immobilized to a planar surface have been used to measure chemically induced damage using an electroluminescence assay format.[10] Incorporation of different cytochrome P450 enzymes into the film spot sensors has facilitated activation and detection of DNA damage caused by pollutants such as benzo[*a*]pyrene.

In addition to the approaches that measure damage to surrogate DNA, derived from calf thymus or salmon sperm, a number of rapid screening assays and biosensors have been reported that are based on *genetically engineered organisms*. These assays respond to chemically induced damage to any part of the genomic DNA that leads to induction of damage-sensitive promoters within the reporter organism. Some of these assay platforms are selectively sensitive to different forms of DNA damage. For example, arrays of bacteria, each engineered to respond to different classes of genotoxic compounds, have been constructed on the surface of glass chips. Differences in their responses to genotoxic compounds have allowed the classification of these compounds into various functional groups.[11] In addition to bacterial systems, yeast-based systems have been used to better approximate damage to the eukaryotic genome. For example, rapid screening assays that use genetically modified yeast have compared favorably with the Ames test and rodent tumorigenic systems.[12]

DNA may be chemically damaged by a number of different types of compounds with a range of mechanisms. Although many of these chemical changes can be repaired by enzymatic processes in the cell, mutations and chromosomal damage can occur from these chemical changes. There are several types of compound interactions that may result in physical or chemical damage to DNA. These interactions include strand breaks, alkylation, adduct formation, cross-linkers, intercalators, cation coordination, and oxidative damage (Table 6.1).

*Strand breaks* involving one or both strands can be formed by free radical attack or as a result of acidic or enzymatic hydrolysis of phosphodiester bonds. DNA damage resulting from free radicals can be generated by exposure of DNA to ionizing radiation or metal ions such as iron in the presence of $H_2O_2$. Bases may be released by the hydrolysis of *N*-glycosidic bonds caused by exposure to oxidative damage and leading to abasic sites. Bases can also be modified by alkylating agents which may change their pairing characteristics. Examples of DNA-damaging compounds include styrene oxide, which forms covalent adducts with guanine and adenine bases, disrupting the double helical structure.[13] Other examples include methyl methane sulfonate (MMS) and ethyl nitrosourea, which form small adducts.[4] Chemical adducts may also form due to bulky groups (*e.g.*, mitomycin C–guanine adducts). Mitomycin C has also been shown to form interstrand and intrastrand cross-links.

In addition to covalent modification, there are a range of genotoxic compounds that can bind noncovalently. For example, cationic species such as transition metals can bind to the phosphate groups along the backbone as well

**Table 6.1** Types of DNA damage measured by biosensors and screening assays.

| Notation | Type of damage | Mechanisms of damage |
| --- | --- | --- |
| SB | Strand breaks | SBs (single or double strand) may occur due to several types of chemical reactions including radical attack on deoxyribose residues, acidic hydrolysis or enzyme-catalyzed hydrolysis of phosphodiester bonds. This type of damage may result in transitions from supercoiled to relaxed or circular to linear forms as well as lowering strand dissociation temperatures (melting temperatures) |
| ALK | Alkylation | Electrophilic agents may react with base residues to change their chemical structure. This type of damage may also lead to SBs |
| AD | Adduct formation | Addition of bulky adducts such as mytomycin C |
| CL | Cross-linking | Bifunctional cross-linker may form interstrand or intrastrand cross-links |
| INT | Intercalation | Aromatic ring systems may stack between bases |
| CAT | Cationic coordination | Transition metals may form coordination complexes with phosphate groups |
| OX | Oxidative damage | Free radicals, peroxides and photoexcited dyes may result in oxidative damage to bases |

as forming coordination complexes with the bases. The binding of $Ni^{2+}$ and $Cd^{2+}$ to DNA has been shown to cause conformational changes but not oxidative damage, whereas $Pb^{2+}$ interacts with DNA leading to oxidative damage and formation of 2,8-oxo-adenine.[14] Another type of noncovalent interaction results from the intercalation or stacking of condensed aromatic ring systems between the base pairs in the helical structure. Examples of intercalating compounds include dyes such as acridine orange and ethidium bromide,[15] and aromatic amines such as 2-anthramine and 2-naphthyl amine.[16]

Measuring genotoxicity through screening of individual compounds or environmental contamination is an important step in the complex procedure of establishing the carcinogenic potential of these compounds or samples. Chemical modification or noncovalent binding of molecules to DNA as well as damage to DNA replication or repair enzymes and structures may result in mutations, mitotic faults or chromosomal aberrations.[4]

Several types of DNA have been used as surrogates for damage-sensitive optically based biosensors and bioanalytical methods. These include supercoiled plasmid DNA, fish testes DNA, calf thymus (type II), or highly polymerized (type I) DNA, and specific oligonucleotide sequences.[17–19] The most common type of surrogate DNA used in these screening assays for DNA damage is commercially available calf thymus (type II).

The detection mechanism used in specific assays may often dictate the type of DNA used and in some cases may require specified sequences of

**Table 6.2** Summary of recent advances in biosensors and rapid bioanalytical screening assays for DNA damage.

| Transducer type | Type of DNA damage | DNA Surrogate | Detection mechanism | Damaging Agent | Relative sensitivity | Reference |
|---|---|---|---|---|---|---|
| Optical | INT | Calf thymus | Evanescent wave, ethidium bromide probe, competition with intercalator | Anthroquinones | +++ | 15 |
| Optical | SB, OX | Oligonucleotide 38-mer | Fiber-optic, hybridization-based Bodipy probe, | Ionizing radiation, SIN-1 | +++ | 20 |
| Optical | INT | Calf thymus Fish testes | Fiber-optic, DNA damage blocks TO-PRO-3 probe binding | Aromatic amines | +++ | 17 |
| Optical | SB, OX, CL, AD | Calf thymus Plasmid DNA | Melting-annealing analysis, PG probe | UV-radiation, restriction enzymes, genotoxins | ++ | 18 |
| Optical | AD | Calf thymus | Melting-annealing analysis, PG probe | SO | + | 24 |
| Optical | CL, ALK | Calf thymus | Melting-annealing analysis, PG probe | Toxic industrial compounds | + | 25 |
| Optical | AD | DNA adduct | Fluorescence of antibody-DNA adduct | BaP analogs | +++ | 26 |
| Optical | SB | Fish testes Red blood cell | Time-resolve fluorescence, chemical unwinding, PG probe | gamma-radiation | ++ | 22 |
| Optical | SB, AD | Calf thymus | DNA damage inhibits TO probe fluorescence | UV radiation | + | 23 |
| Optical | SB, OX | Oligonucleotide 25-mer | DNA damage inhibits binding of cationic polymer to 25mer | Fenton Rxn, S1 nuclease | ++ | 21 |

# Nucleic Acid-based Biosensors for the Detection of DNA Damage

| Method | Damage | Recognition | Description | Analyte | Rating | Ref |
|---|---|---|---|---|---|---|
| Optical | SB, OX | Oligonucleotide Stem-loop | resulting in a red shifted absorbance spectra | Bleomycin, Fenton rxn | +++ | 19 |
| Electro-chemiluminescence | AD | Calf thymus | Fluorescence Resonance Energy Transfer detection, "molecular breaklights" | BaP | ++ | 10 |
| Electro-chemiluminescence | AD | Calf thymus | CV luminescence, PG electrode-metalopolymer film [Ru(bpy)$_2$PVP$_{10}$]$^{2+}$, enzymes CYT P450 | Styrene | + | 27 |
| Acoustic | OX | Oligonucleotide 19 mer | CV luminescence-CCD camera, PG electrode-metalopolymer film [Ru(bpy)$_2$PVP$_{10}$]$^{2+}$, enzyme CYT P450 | Thymine glycol integrated to simulate oxidative damage | + | 28 |
| Acoustic | OX | Oligonucleotide 19 mer | Acoustic wave, Δ-frequency, Δ-motional resistance | Apuric or apyridimic sites were inserted to simulate oxidative damage | + | 29 |
| Fluorescence Flow Cytometry | ALK, SB | Yeast | Acoustic wave, Δ-frequency, Δ-motional resistance | MME, MMS, camptothecin | + | 30 |
| Chronoamperometric | AD | Bacteria | Yeast HUG1P::GFP promoter-reporter | 4-NQO, IQ | + | 31 |
| Fluorescence flow-through cell | ALK, AD, CL | Bacteria | Bacterial SOS-response, pre-treatment with S9 liver microsomes | MMC | +++ | 32 |
| Bioluminescence immobilized to 96 well plate | ALK, AD, CL | Bacteria | Bacterial recA::mCherry, recA::gfpmut 3.1 | MMC | +++ | 33 |
| | | | Bacterial lux fusions, 12 strains | | | |

**Table 6.2** (continued)

| Transducer type | Type of DNA damage | DNA Surrogate | Detection mechanism | Damaging Agent | Relative sensitivity | Reference |
|---|---|---|---|---|---|---|
| Fluorescence single cell format | ALK | Yeast | Yeast RAD54-GFP S9 liver microsomes | MMS | +++ | 12 |
| Bioluminescence 96 well plate format | SB, AD, ALK,OX, CL | Bacteria | Bacterial nrdA::luxCDABE | NDA, MMC, MNNG, 4-NQO, $H_2O_2$ | ++ | 34 |
| Bioluminescence 96 well plate format | SB, AD, ALK, CL | Bacteria | Bacterial ADP1_recA::lux | BaP, MMC, MMS, UV | +++ | 35 |
| Bioluminescence continuous flow cell | ALK, AD, OX, CL | Bacteria | Bacterial lux fusions, 4 strains | MMC, paraquat, $H_2O_2$ | +++ | 36 |
| Bioluminescent micro-chip array | OX | Bacteria | Bacterial lux fusions, 12 stains | Paraquat analogs, $H_2O_2$ | +++ | 11 |

Abbreviations: SB: strand breaks, ALK: alkylation damage, AD: adduct formation, CL: cross-linker damage, INT: intercalator damage, CAT: cationic coordination, OX: oxidative damage, PG: PicoGreen, TO: thiazole orange, TP3: TO-PRO-3, CV: cyclic voltammetry, CYT: cytochrome, SO: styrene oxide, BaP: benzo[a]pyrene, SIN-1: 3-morpholinosydnonimine, NDA: nalidixic acid, MMC: mitomycin C, EMS: ethane methylsulfonate, MMS: methyl methanesulfonate, MNNG: 1-methyl-1-nitroso-N-methylguanidine, 4-NQO: 4-nitroquinoline N-oxide, IQ: 2-amino-3-methylimidazo[4,5f]quinoline.

oligonucleotide. For example, several assays for DNA damage have been reported that use hybridization-based formats including complementary target and capture oligonucleotides,[20] formation of stem-loop structures,[19] or binding of a cationic polythiophene probe to a single-stranded (ss) oligonucleotide[21]. Other assays detect damage to genomic DNA and require no specific sequence, such as those employing calf thymus DNA, fish testes DNA, or DNA isolated from red blood cells.[22]

## 6.2 Optical Transduction Schemes

The particular type of biosensor or screening assay used to measure DNA damage will significantly influence the strategy used to detect the resulting physical or chemical damage. Transduction schemes used for optically based screening assays that measure damage to surrogate as well as microbial DNA depend on the type of nucleic acid surrogate and the type of damage being measured (Table 6.2).

Mechanisms for detection of DNA damage using optically based screening assays also depend, to some extent, on the detailed characterization and commercial availability of a number of dyes with high affinity for chemical and structural features of DNA. More specifically, cyanine dyes such as TO-PRO-3 (TP3), SYBR GREEN (SG), thiazole orange (TO), and PicoGreen (PG) show significantly enhanced fluorescence as a result of intercalation into double-stranded (ds) DNA, allowing them to be used as fluorescent probes.[37,38] Detailed studies of the fluorescence response at various dye/base pair ratios has suggested that SG binds to dsDNA through both intercalation and external binding.[38] These investigators also suggested that the increase in fluorescence resulted from external binding. For PG, it has been proposed that when this dye is free in solution, it can dissipate excitation energy by rotation around its central methine bridge, whereas when it is intercalated into dsDNA it is rotationally restricted and shows a large increase in observed fluorescence due to an increase in quantum yield.[39]

Several types of screening assays for DNA damage take advantage of structural changes that are sensitive to the unwinding process. One optical screening approach for detection of nucleic acid damage involves observation of both the temperature-induced dissociation (unwinding) and the process of reannealing of DNA.[18,24,25] Surrogate DNA, used for biosensors and rapid screening assays, often becomes more susceptible to unwinding after it has been chemically modified (damaged) by a wide range of compounds or by ionizing or UV radiation. The basis for this approach is in some ways similar to the single cell electrophoresis (comet) assay, the alkaline unwinding assay, and the alkaline elution assay.[2] In addition to differences in the damage-dependent dissociation (melting) of DNA, the re-association (annealing) of DNA is also sensitive to various forms of damage (Figure 6.1).

This process can be measured in real time using the dsDNA selective dye PG. The melting/annealing assay for DNA damage has been demonstrated using

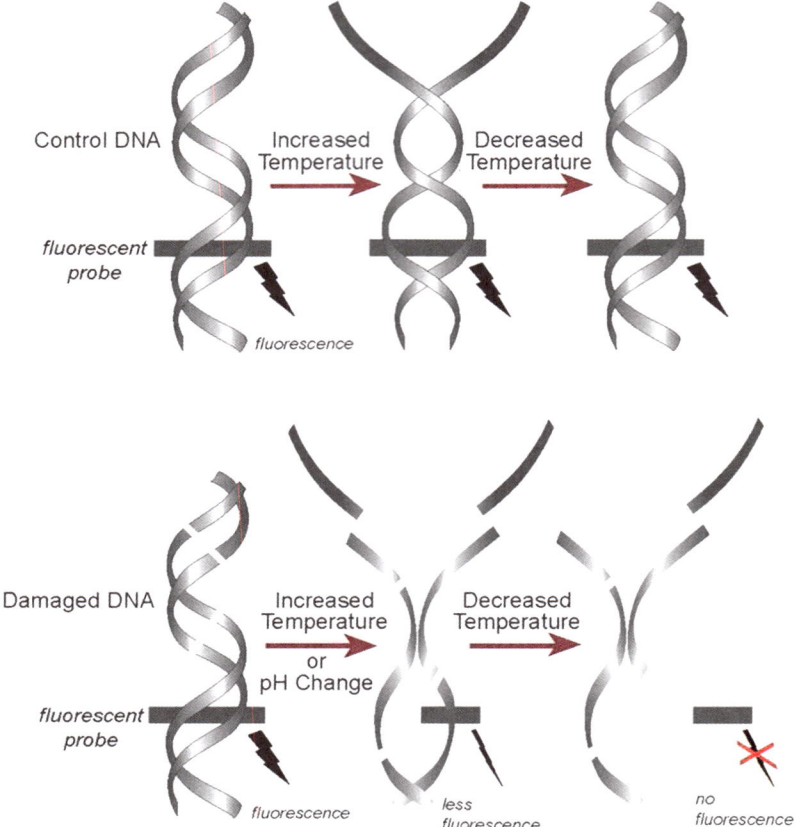

**Figure 6.1** Fluorescence response of control (undamaged) DNA or damaged DNA resulting from temperature (melting) or pH-induced denaturation and subsequent renaturation (annealing) using a fluorescent indicator dye such as PicoGreen (PG).

specific restriction enzymes that selectively cut plasmid (pUC19) DNA. For this assay, an increased number of cuts in the plasmid DNA corresponded to less of the dsDNA after each melting/annealing cycle.[18] The melting/annealing process was also sensitive to other types of DNA damage.

For example, exposure of plasmid DNA to DNA-damaging compounds including glutaraldehyde, methanesulfonic acid methyl ester, pyrene, benzo[a]pyrene, 4-nitroquinoline-N-oxide, 1,3-butadiene diepoxide, and styrene oxide significantly inhibited reannealing of thermally denatured plasmid DNA.[18] This assay format was also shown to detect damage resulting from ionizing radiation,[39] UV radiation, alkylation, chemical oxidation, and exposure to cross-linkers, intercalators, and adduct-forming compounds.[18,24,25]

Approaches using competitive intercalation have been reported for detection of genotoxic polycyclic aromatic hydrocarbons (PAHs).[15,17] Important design

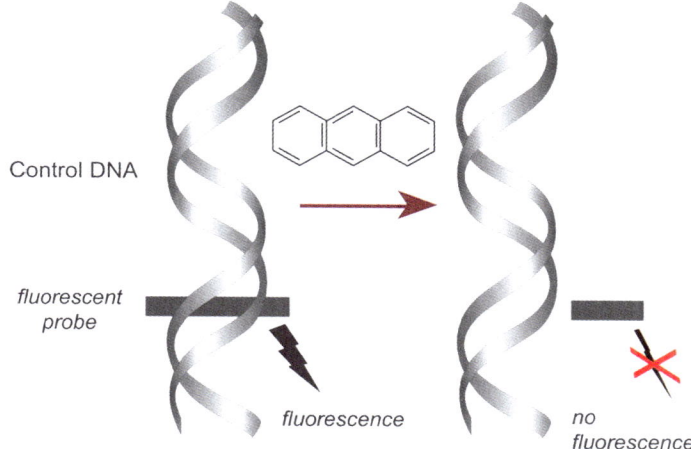

**Figure 6.2** Decrease in fluorescence resulting from competitive displacement of the fluorescent indicator dye (probe) by genotoxic intercalator. Fluorescent dyes such as PG, TO-PRO-3, or SYBR GREEN exhibit maximal fluorescence when associated with dsDNA.

features for the rapid screening assay reported by Liu and Danielsson[17] included the use of a well-defined nucleic acid surrogate provided by restriction endonuclease digested plasmid (pBR322) DNA and the choice of TP3 as the reporter dye. TP3, as opposed to the higher-affinity dye SYBR green, was more easily displaced by competitive intercalators yielding a more sensitive assay (Figure 6.2). This fiber-optic sensor system detected a range of toxic aromatic amines (2-naphthylamine, 2-anthramine) in the nanomolar concentration range.[17]

In a similar assay format, the DNA-staining dye TO was used as the basis for a rapid screening assay for DNA damaged by UV radiation.[23] This assay measured the damage-dependent decrease in fluorescence for the DNA-intercalating dye TO. Although the assay response required relatively high levels of UV radiation, the change in fluorescence after DNA damage was greater than observed changes in absorbance, circular dichroism, or agarose gel electrophoresis migration.[23]

In addition to differences in fluorescence intensity between ssDNA–dye complexes and dsDNA–dye complexes, dyes such as PG also exhibit differences in characteristic fluorescence lifetimes[22] when bound to these forms of DNA. Using the measurement of fluorescence lifetimes, a rapid and sensitive optical screening assay for DNA damage has been reported for red blood cell-DNA exposed to $\gamma$-radiation.[22] After the radiation-exposed cells were lysed, the damaged DNA was placed in buffer that selectively dissociated (unwound) the nicked strands. The dissociated DNA strands were then stained using PG. The time-resolved fluorescence measurements were used to differentiate the relative percentage of dsDNA and ssDNA, without the need to separate each species, and independent of the total amount of DNA present.[22]

In a unique assay format, a stem–loop oligonucleotide sequence was used as surrogate DNA to construct a rapid and continuous assay for strand cleavage.[19] This assay format, termed "break lights," used fluorescence resonance energy transfer (FRET) with fluorescein as the 5′-fluorophore and 4-(4′-dimethylaminophenylazo) benzoic acid as the 3′-quencher. The DNA-damaging compounds bleomycin and methidiumpropyl-EDTA-Fe(II) were observed at nanomolar concentration levels. Advantages for this assay included speed, sensitivity, and simplicity in both reagents required for the assay and instrumentation (visible spectrophotometer).

Another simple and rapid assay format for detection of DNA damage has been reported using a cationic polythiophene derivative[21] that shifted the absorbance when bound to ssDNA. The polythiophene indicator shows a red-shifted absorbance spectrum when bound to a ss 25-mer oligonucleotide. When the oligonucleotide was cleaved by Fenton-reaction generated hydroxyl radicals or DNA nuclease, the indicator molecule reverted to its random-coil configuration resulting in a visible absorbance change (red to yellow). Due to this visible color change, the assay endpoint can be detected by UV-visible spectrum or by noninstrumental observation.[21]

In another optical screening assay format, fluorescent benzo[a]pyrene-derived DNA adducts were detected using a monoclonal antibody-modified gold biochip. Low temperature (77 K) laser-induced fluorescence spectra were used to measure a benzo[a]pyrene-G-$N^7$ guanine adduct present in the damaged DNA.[26] Although this assay required relatively complex processes and instrumentation, it was specific and sensitive (femtomole detection limits).

Similar to the electrode biosensors that measure DNA damage using $[Ru(bpy)_3]^{2+}$ as a catalyst to report guanine oxidation, a reported detection mechanism for electroluminescent/voltammetric screening assays for DNA damage depends on the use of the metallopolymer $[Ru(bpy)_2PVP_{10}]^{2+}$ as a probe. For this assay, DNA was immobilized in a thin film also containing cytochrome P450. Upon exposure to styrene, both the square wave voltammetry (SWV) signal and the electroluminescence signal increased.[27] The proposed mechanism for increased guanine oxidation was related to increased access due to disruption in the helical structure of the damaged DNA. This assay format was also configured into a sensor array and used to measure the efficacy of isoforms of cytochrome P450 in the metabolic activation and subsequent DNA damage caused by benzo[a]pyrene.[10]

## 6.3 Whole-Cell Based Biosensors

Live cells can be used to quantify genotoxins in various ways. The modern concept of the whole cell biosensor requires a system that delivers rapid readout in a convenient, self-contained, and portable format. For many of the recently described microbial screening assays for DNA damage, the organisms are not directly interfaced to an optical transducer; however, these assay formats can be miniaturized, automated, and potentially configured to be field portable.

Traditional mutagenicity assays cannot satisfy these specifications, but they certainly illustrate the utility of employing live organisms as biological detectors to measure the relative concentrations of mutagens and, by inference, genotoxins. A well-known example of a mutagenicity assay is the Ames test (Figure 6.3). This test uses special strains of the bacterium *Salmonella typhimurium* that contain nucleotide substitution or frameshift mutations in genes required for histidine biosynthesis. The cells also possess defects in endogenous DNA repair pathways and other traits that enhance sensitivity to mutagens. Histidine auxotrophy serves as an easily assayed genetic marker; these bacterial strains can only grow on a plate containing histidine-deficient media if a reversion mutation occurs that restores the ability to make histidine via biosynthesis. When these cells are exposed to genotoxins, DNA damage occurs at a frequency that is roughly proportional to the concentration of the DNA damaging agent. During DNA replication and cell division, unrepaired DNA lesions can give rise to heritable mutations that become fixed in the progeny. If the mutation occurs in the *his* (histidine biosynthesis) gene in such as way as to restore gene function, then a visible colony will be produced after a period of growth on a histidine-minus plate. Mutations that occur elsewhere in the genome are not observable. Because each revertant colony represents a mutational event that occurred in a progenitor cell, simply counting the colonies gives an estimate of the relative mutation rate elicited by treatment with the test agent. The Ames assay measures mutagenicity, and not genotoxicity *per se*, but there is in general a cause–effect relationship between DNA damage and mutation, so it is reasonable to infer genotoxicity from Ames test results in most cases.

The biosensor concept that is the focus of this review is fundamentally different from any sort of mutagenicity assay in several respects (Figure 6.3). First of all, whole-cell-based biosensors measure genotoxicity, not mutagenicity. These systems do not require that cell division takes place in order to create an observable output signal. Thus, genotoxin quantification can proceed on a timescale of minutes, consistent with the biosensor concept, not the days needed for mutagenesis assays. Whole-cell biosensors have the potential to be more sensitive than mutagenesis assays, because the DNA target is typically orders of magnitude larger. In a mutagenesis assay, only DNA lesions that occur within the sequence of the genetic marker (*e.g.*, the *his* gene, in the case of an Ames test) can be detected. In contrast, DNA lesions that occur anywhere within the organismal genome can initiate a response in a biosensor system. In addition, endogenous DNA damage response pathways (*e.g.*, the bacterial SOS/*recA* signaling network) serve as biochemical amplifiers that further enhance sensitivity in many assay types. A final distinction lies in the direct coupling of signal generation and signal detection. Unlike mutagenesis assays, which are labor-intensive procedures, whole cell biosensors measure a real-time optical or amperometric signal which is proportional to the amount of DNA damage; these types of readouts can be instantly and automatically quantified without the involvement of the operator of the biosensor device.

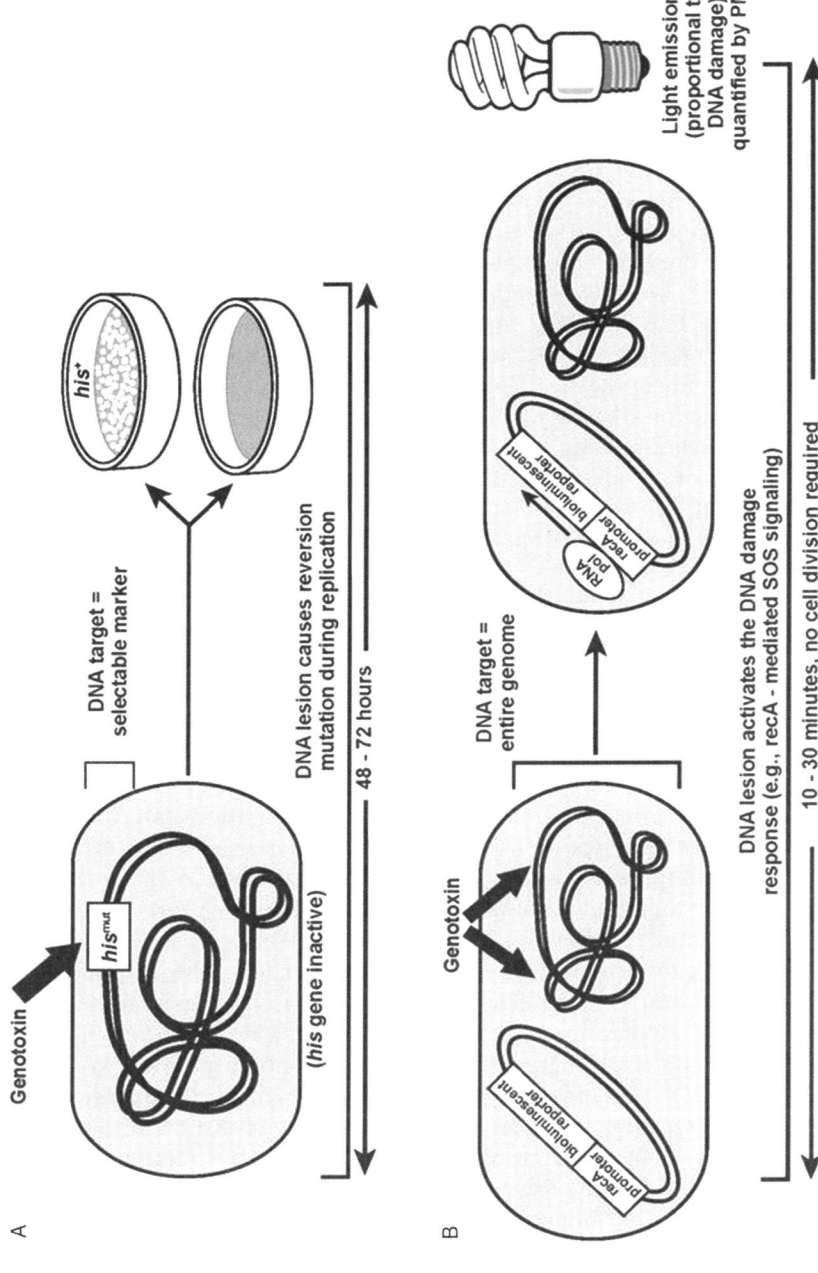

**Figure 6.3** Comparison of mutagenesis assay (Ames test) concept (A) *versus* a genotoxin biosensor concept (B), exemplified by a bacterial strain genetically engineered to carry a reporter plasmid with the *recA* promoter.

This section will examine the various types of cell-based biosensors that are under study, with primary emphasis on work that has been conducted after the most recent major reviews of this field.[6,40] Bacteria, yeast, and even mammalian cells have been used as sensors in biosensor prototypes. In each case, the sensor function is based upon the cell's innate ability to recognize DNA damage to its own genome and then mount a transcriptional response. Bacteria are inexpensive to culture and easy to manipulate genetically, so the majority of whole-cell biosensor development efforts have focused on these simple prokaryotes. DNA damage elicits the bacterial SOS response, a coordinated up-regulation of DNA repair proteins which is essential for cell survival after genotoxin exposure. The SOS response is ubiquitous in bacterial species, but it has been studied most extensively in *E. coli*. The key regulatory proteins of the response are LexA, a transcriptional repressor, and RecA, the actual sensor of DNA damage. The latent coprotease activity of RecA becomes activated when it is complexed with ssDNA, which arises wherever DNA lesions block the progression of DNA polymerase during replication[41] and near sites of dsDNA breaks whose ends have been unwound by bacterial helicase enzymes.[42] The activated RecA coprotease causes proteolytic degradation of LexA, which relieves transcriptional repression by LexA at various SOS-responsive promoters. The *recA* gene is itself controlled by such a promoter, in an autoregulatory positive feedback loop that rapidly amplifies the cellular response to DNA damage. Altogether, including the genes for RecA, the nucleotide excision repair proteins, recombination proteins, and others, there are at least 30 or so distinct SOS-regulated genes.[43] The promoter sequence from any of these genes could in theory be coupled to a suitable reporter gene to create a recombinant transcriptional unit that functions as the core of a cell-based biosensor, but obviously the specific choices of promoter and reporter will greatly affect the performance of the system. The rapid kinetics of the SOS response make it viable for adaptation to biosensor technology; when *E. coli* cells are exposed to $5\,J\,m^{-2}$ of UV irradiation, the LexA repressor protein pool is about 80% degraded within 3 min, and 90% degraded at 5 min.[41] Following derepression, transcription and translation of reporter protein begins immediately, and appreciable (though not maximal) levels of reporter protein can be expected to begin to accumulate in as little as 10 min (Figure 6.3). If rapid response time is a priority, the specific type of reporter protein must be taken into account. Whereas most enzymatic reporters, such as lux or b-gal, are active immediately upon biosynthesis, spontaneous post-translational reaction steps are needed for maturation of the fluorophore in green fluorescent protein (GFP). Depending on the particular GFP variant, time constants ranging from 30–120 min have been reported for the development of the active fluorophore.[44]

Different SOS-responsive promoters have differing affinities for LexA repressor protein, and therefore can be expected to vary in sensitivity and timing of responsiveness to genotoxic stimuli. Plasmids were constructed to directly compare four distinct promoters from *E. coli* (*cda*, *recA*, *sulA*, and *umuDC*) as drivers for a GFP reporter.[45] Of these four, the *cda* promoter gave the largest magnitude GFP induction in response to 1-methyl-1-nitroso-$N$-methylguanidine (MNNG), with the *cda* promoter being 12 times more responsive than the *recA* promoter. In this system, the detection limit for the genotoxins MNNG and mitomycin C were 0.16 μM and 9 nM, respectively. In a subsequent study by the same group, the *cda*/GFP strain was applied to model soil samples containing mitomycin C.[46] After 24 h of residence time in the soil, the bacterial fraction was recovered by Nycodenz gradient centrifugation and analyzed by flow cytometry, and the minimum detectable concentration of soil contaminant was found to be 2.5 ng mitomycin C per gram of soil. A flow cytometer is an expensive laboratory instrument; as an alternative, Martineau *et al.* fabricated a small flow cell with light emitting diode (LED) excitation sources to evaluate as a portable biosensor device for field use.[32] This unit used *E. coli* harboring the *recA*/GFP promoter/reporter combination, as well as a similar construct substituting the red fluorescent protein mCherry in place of GFP. With this apparatus, the authors reported limits of detection for mitomycin C of 115 nM and 12 nM for *recA*/GFP and *recA*/mCherry, respectively, based on twofold fluorescence above background. If a signal-to-noise ratio of 3 or better is taken as the criterion for detection, the limits reported in this study are lowered to 2.0 nM and 0.25 nM, respectively.

The research group of Man Bock Gu has published extensively on the use of *E. coli* as a biosensor for detection of DNA-damaging agents as well as other categories of toxins. They have created and characterized a large number of promoter/reporter combinations, some of which contain a *recA* promoter and are designed to sense genotoxins, and others that employ oxidative stress-specific promoters that are designed to sense agents that generate reactive oxygen species. Of course, reactive oxygen species such as superoxide anion and hydroxyl radical cause DNA damage, but these agents attack proteins and lipids as well. Because the physiological defense responses elicited by this class of agent differ from that of the SOS response, it is possible to create biosensors that differentiate among these agents as well. Gu's group has used methyl viologen (paraquat) as a model compound to generate superoxide anion, and hydrogen peroxide as the source of hydroxyl radicals.[11,33,47] Superoxide preferentially activates the *sodA* and *pqi-5* promoters, whereas peroxide preferentially activates the *katG* promoter.[11,33,47] Treatment with $H_2O_2$ also stimulates the *recA* promoter, presumably due to the oxidative DNA damage it causes, although the *recA* promoter is not as sensitive to $H_2O_2$ as the *katG*.[33] Instead of GFP, this group has used Lux proteins, which produce bioluminescence *in situ* without requiring the addition of any exogenous substrate reagents. In 2004, this group directly compared reporter constructs expressing bioluminescent proteins produced from the *lux* operons of either the more thermolabile *Vibrio fischeri* or the more thermostable *Xenorhabdus luminescens* (now called *Photorhabdus luminescens*), and concluded that the latter generally performed somewhat better in most systems.[47]

One of the major objectives of Gu's group has been to utilize a broad panel of biosensor bacterial strains in an array format that enables one to categorize toxicological mode of action. Lee *et al.* used 20 different *E. coli* strains, including those with recA/Lux, sodA/Lux, pqi-5/Lux, and katG/Lux promotor/reporter combinations, arrayed in a 384-well plate and challenged with either mitomycin C, paraquat, or salicylic acid, and demonstrated differential responses as a result of DNA damage, oxidative damage, or membrane damage in response to these agents.[48] A similar array format was characterized subsequently by this group in 2007,[11] but this study focused exclusively on oxidative stress agents such as $H_2O_2$ and did not aim to quantify DNA damage *per se*. Mitchell and Gu looked to test whether the various *E. coli* biosensor strains could be immobilized using agar or sol-gel in the array format.[33] For the recA/Lux strain, the lowest detectable concentration, defined as that eliciting a twofold increase in bioluminescence compared to untreated control, for mitomycin C, was 0.0006 mg/L in liquid culture and 0.001 mg/L for the immobilized format. The corresponding sensitivities to hydrogen peroxide for this strain were 1.5 mg/L and 11.7 mg/L, respectively. Lee and Gu have also developed a prototype unit that uses some of these biosensor strains in a two chamber mini-bioreactor set-up that would enable continuous water quality monitoring.[36]

Recently, Gu's group has developed a new plasmid construct for DNA damage biosensors. The *E. coli* nrdA promoter is expressed in response to various types of DNA-damaging agents, but unlike the recA promoter, it is not dependent on repression by LexA. The *E. coli* strain with nrdA/Lux responded to the genotoxins MNNG, mitomycin C, nalidixic acid, and 4-NQO, but did not respond to paraquat or various phenolic compounds.[34] In this system, the minimum detectable concentration for mitomycin C was 0.3125 ppm. Hydrogen peroxide, a less potent genotoxin, was detected down to 50 ppm. This work has been extended to include yet another class of promoter responsive to DNA damage, alkA, which responds specifically to DNA-alkylating agents.[49] Various types of genotoxins produce differential responses in reporter systems driven by recA, nrdA, and alkA promoters, indicating that it may be possible to classify mode of action of genotoxic compounds using bacterial biosensor arrays.[49]

Song *et al.*[35] have integrated recA/Lux into *Acinetobacter baylyi* instead of *E. coli*, to produce a bacterial cell biosensor strain that remains viable under harsher conditions such as prolonged storage at 4 °C. This strain responded to mitomycin C with a detection limit of 1.5 nM in a 3-h assay. Many potential mutagens require metabolic activation, and this group demonstrated that the *A. baylyi* strain was compatible with microsomal S9 conversion for the detection of benzo[*a*]pyrene down to 0.4 nM. Buchinger *et al.* employed the *Salmonella typhimurium* bacteria with a SOS response promoter coupled to β-galactosidase as a reporter enzyme.[31] This reporter activity was quantified using both *o*-nitrophenyl-β-D-galactopyranoside, a colorimetric substrate whose product *o*-nitrophenol absorbs at 414 nm, and *p*-aminophenyl-β-D-galactopyranoside, an amperometric substrate whose product *p*-aminophenol can be oxidized electrochemically within a 200–400 mV potential window with minimal interference from other redox-active species. The group demonstrated that the system is compatible with the S9 mix used for metabolic activation, and they demonstrated excellent correlation between colorimetric and amperometric signals elicited by the genotoxic compound IQ (2-amino-3-methylimidazo[4,5-f]quinoline) over the dose range of 0.16–5.0 μM. This work raises the hope that a simple, inexpensive electrode-based biosensor could function in place of optical detection.

We conclude with a look at *eukaryotic cell systems* that have been examined as biosensors. In yeast, DNA damage causes expression of a group of DNA repair proteins. The promoter of one these, *RAD54*, has been coupled to a GFP reporter. Recently, a very broad survey of genotoxic and potentially genotoxic compounds was completed (305 distinct compounds in all), and the *RAD54*/GFP results were systematically correlated with results from more cumbersome assays such as the Ames test and micronucleus formation assay, with generally good agreement.[12] Alternatively, a yeast *HUG1*/GFP promoter/reporter construct has been used to quantify DNA damage produced by various agents, using fluorescence as measured by flow cytometry.[30] In this system, additional mutations were engineered into the strain that knocked out base excision repair and ds break repair pathways to increase sensitivity of the yeast biosensor to certain classes of genotoxin.

*Human cells* represent an interesting choice as the basis for a whole-cell-based biosensor. On the one hand, the effect of potential genotoxins on human cells is of much greater interest from a public health perspective, compared to the responses of bacteria or yeast. On the other hand, whereas bacteria and yeast are easily stored, transferred, and cultured, it is unlikely that the much more delicate human cells could ever be adapted into a portable biosensor device for use in the field. Nonetheless, in a laboratory setting, human cells could be used in high-throughput testing to evaluate genotoxicities. The human transcription factor p53 is strongly activated in response to almost any type of DNA damage. A recent study has systematically compared the performance of three commercially available screening kits (GreenScreen, CellCiphr, and CellSensor) that employ some type of p53-dependent response coupled to a reporter.[5] Out of 320 compounds evaluated, 71 compounds were scored as positive for genotoxicity by at least one of the three assays, but of these only 22 were positive in two or all three of the assays, indicating that a degree of nonconcordance exists in the overall characterization of compounds using these various systems.

## 6.4 Field Capability and Stage of Development

Potential advantages offered by optically based biosensors and rapid screening assays with respect to environmental monitoring for genotoxic compounds involves potential miniaturization (and implied portability and reduced manufacturing cost), speed (in the minutes to hours window), simplicity (for execution and interpretation of data), and the applicability to complex matrices with a minimum of sample preparation. With respect to field applications, optical instruments that employ fiber optics, plate readers, and microchip formats have the potential to be miniaturized into field-portable devices. These technological improvements also apply to acoustic devices and instrumentation. In addition, immobilization of surrogate DNA in stable forms to the surface of microchips or microwells, as well as the use of automated pumps and microfluidics, can be used to adapt many of these techniques for potential field application.

The detection of biochemical changes that occur as a result of DNA damage in genetically engineered microorganisms has been facilitated primarily using fluorescence[12,30,32] and bioluminescence techniques.[11,33–36] Instrumentation for each of these detection modalities can potentially be adapted for field applications. In addition, significant advances have been made toward lyophilizing and stabilizing these organisms for potential use-on-demand formats[35] or continuous-use bioreactors.[36]

## 6.5 Future Trends and Summary

Future trends for the development and improvement of biosensors and rapid bioanlytical screening assays for DNA damage are likely to be in the following areas: (1) incremental improvements in previously reported techniques with respect to sensitivity, speed, and high-throughput formats; (2) increased versatility in the range of compounds (with various mechanisms of action) for which assays will respond; and (3) demonstration of assays using environmental matrices and under field conditions. Technological innovations that may drive these improvements include microfluidics, sensor arrays, and nanotechnology (materials and structures). For example, the hybridization mechanism is (directly or indirectly) the basis for a range of biosensor and bioanalytical screening assays discussed in this chapter.[18–20,28,29] Hybridization assays using oligonucleotide sequences for detection of single-base mismatches have been significantly improved for both electrochemical[50] and optical[51] detection using nanomaterials. Similarly, nanomaterial-enabled techniques could significantly impact biosensors and rapid bioanalytical screening assays for DNA damage.

Part of the value for the use of bioassays that measure chemically induced DNA damage rather than measuring the individual compounds themselves is the ability to screen for the presence of numerous potential genotoxins. The characterization of biosensors and rapid screening assays for a broad range of genotoxins increases their potential versatility. One of the major challenges for environmental screening and monitoring assays is the variable matrices and the ability to develop and demonstrate these assays under field conditions (*e.g.*, variations in temperature, light, wind, and meeting power requirements). Future trends for environmental applications will likely include more reports of field demonstrations.

In summary, there are a wide range of devices and techniques that have been developed and demonstrated for rapidly detecting the effects of genotoxins on surrogate or cellular DNA. Although these techniques range along a continuum from electrochemical biosensors through optically based screening assays that detect damage to surrogate DNA to genetically engineered microorganisms, many of them show the potential to be developed for environmental field screening and monitoring applications. There are, however, some issues that should be considered when viewed in the context of environmental monitoring requirements. First, responses for reported biosensors and screening assays are typically described for a limited number of compounds that exert damage to DNA by means of specific mechanisms (*e.g.*, strand breaks, alkylation, adduct formation, cross-linking, intercalation, cation binding, or oxidative damage). Consequently, the responses of these optical and bacterial assays to the wide range of genotoxins that contaminate the environment have largely been unexplored. For the most part, these assays have been reported in controlled buffer solutions and under laboratory conditions. It is encouraging to note, however, that in cases where electrochemical biosensors that measure damage to surrogate DNA were used for analysis of genotoxins in environmental water samples, only simple filtration steps were required.[16,52,53]

Another obstacle for putting optically based rapid screening assays for DNA damage into the hands of potential environmental users is *commercialization*. Commercialization of these assays for environmentally relevant applications faces significant hurdles. Some of these obstacles include the diversity of samples, the variability of data quality requirements among environmental programs, the broad range of environmental monitoring applications, and the reluctance of regulators to adopt toxicity and genotoxicity endpoints into local and national regulations. As a result, development cycles are likely to be long and commercial success will continue to remain limited.[54,55] Nevertheless, there are a number of areas where the unique capabilities of rapid screening assays might be exploited to meet the requirements of environmental monitoring for genotoxic compounds. Continued advances in the areas of nanomaterial-enabled screening assays as well as microfluidics and automated operation could widen the potential market and allow these techniques to be adopted for routine use.

## Disclaimer

The United States Environmental Protection Agency (EPA), through its Office of Research and Development (ORD), has funded and managed the research described here. It has been subjected to the Agency's administrative review and has been approved for publication. Mention of trade names or commercial products does not constitute endorsement or recommendation for use.

## References

1. M. Farre and D. Barcelo, *Hdb Env. Chem.*, 2009, **5J**, 115.
2. P. D. Hansen, E. Wittekindt, J. Sherry, E. Unrun, H. Dizer, H. Tug, H. Rosenthal, V. Dethlefsen and H. von Westernhasen, *Hdb Env. Chem.*, 2009, **5J**, 203.

3. I. Palchetti and M. Mascini, *Analyst*, 2008, **133**, 846.
4. A. Stang and I. Witte, *Mut. Res.*, 2009, **675**, 5.
5. A. W. Knight, S. Little, K. Houck, D. Dix, R. Judson, A. Richard, N. McCarroll, G. Akerman, C. Yang, L. Birrell and R. M. Walmsley, *Regul. Toxicol. Pharmacol.*, 2009, **55**, 188.
6. K. R. Rogers, *Anal. Chim. Acta*, 2006, **568**, 222.
7. A. R. Collins, *Am. J. Clin. Nutr.*, 2005, **81**, 261S.
8. G. Bagni, S. Hernandez, M. Mascini, E. Sturchio, P. Boccia and S. Marconi, *Sensors*, 2005, **5**, 394.
9. M. Badihi-Mossberg, V. Buchner and J. Rishpon, *Electroanalysis*, 2007, **19**, 2015.
10. E. G. Hvastkovs, M. So, S. Krishnan, B. Bajrami, M. Tarun, I. Jansson, J. B. Achenkman and J. F. Rusling, *Anal. Chem.*, 2007, **79**, 1897.
11. J. H. Lee, C. H. Young, B. C. Kim and M. B. Gu, *Biosens. Bioelectron.*, 2007, **22**, 2223.
12. A. W. Knight, N. Billinton, P. A. Cahill, A. Scott, J. S. Harvey, K. J. Roberts, D. J. Tweats, P. O. Keenan and R. M. Walmsley, *Mutagenesis*, 2007, **22**, 409.
13. B. Wang, I. Jansson, J. B. Schenkman and J. F. Rusling, *Anal. Chem.*, 2005, **77**, 1367.
14. S. C. B. Oliveira, O. Corduneanu and A. M. Oliveira-Brett, *Bioelectrochem.*, 2008, **72**, 53.
15. P. C. Pandey and H. H. Weetall, *Anal. Chem.*, 1995, **67**, 787.
16. G. C. Chiti, G. Marrazza and M. Mascini, *Anal. Chim. Acta*, 2001, **427**, 155.
17. Y. Liu and B. Danielsson, *Anal. Chem.*, 2005, **77**, 2450.
18. K. Ramanathan, R. K. Gary, A. Apostol and K. R. Rogers, *Curr. Appl. Physics*, 2003, **3**, 99.
19. J. B. Biggins, J. R. Prudent, D. J. Marshall, M. Ruppen and J. S. Thorson, *Proc. Natl. Acad Sci. U. S. A.*, 2000, **97**, 13537.
20. K. R. Rogers, A. Apostol, S. J. Madsen and C. W. Spencer, *Anal. Chim. Acta*, 2001, **444**, 51.
21. Y. Tang, F. Feng, F. He, S. Wang, Y. Li and D. Zhu, *J. Am. Chem. Soc.*, 2006, **128**, 14972.
22. G. Cosa, A. L. Vinette, J. R. N. McMean and J. C. Scaiano, *Anal. Chem.*, 2002, **74**, 6163.
23. C. C. Trevithick-Sutton, L. Mikelsons, V. Filippenko and J. C. Scaiano, *Photochem. Photobiol.*, 2007, **83**, 556.
24. K. Ramanathan and K. R. Rogers, *Sens. Actuat. B*, 2003, **91**, 205.
25. S. Kailasam and K. R. Rogers, *Chemosphere*, 2007, **66**, 165.
26. N. M. Grubor, R. Shinar, R. Jankowiak, M. D. Porter and G. J. Small, *Biosens. Bioelectron.*, 2004, **19**, 547.
27. M. So, E. G. Hvastkovs, S. B. Schenkman and J. F. Rusling, *Biosens. Bioelectron.*, 2007, **23**, 492.
28. F. Yang, E. Romanova, E. Kubareva, N. Dolinnaya, V. Gajdos, O. Burenina, J. S. Ellis, T. Oretskaya, T. Hianik and M. Thompson, *Analyst*, 2009, **134**, 41.

29. T. Hianik, X. Wang, S. Andreev, N. Dolinnaya, T. Oretskaya and M. Thompson, *Analyst*, 2006, **131**, 1161.
30. M. G. Benton, N. R. Glasser and S. P. Palecek, *Biosens. Bioelectron.*, 2008, **24**, 736.
31. S. Buchinger, P. Grill, V. Morosow, H. Ben-Yoav, Y. Shacham-Diamand, A. Biran, R. Pedahzur, S. Belkin and G. Reifferscheid, *Anal. Chim. Acta*, 2010, **659**, 122.
32. R. L. Martineau, V. Stout and B. C. Towe, *Biosens. Bioelectron.*, 2009, **25**, 759.
33. R. J. Mitchell and M. B. Gu, *Biosens. Bioelectron.*, 2006, **22**, 192.
34. E. T. Hwang, J. M. Ahn, B. C. Kim and M. B. Gu, *Sensors*, 2008, **8**, 1297.
35. Y. Song, G. Li, S. F. Thornton, I. P. Thompson, S. A. Banwart, D. N. Lerner and W. E. Huang, *Environ. Sci. Technol.*, 2009, **43**, 7931.
36. J. H. Lee and M. B. Gu, *Biosens Bioelectron.*, 2005, **20**, 1744.
37. X. Yan, R. C. Habbersett, T. M. Yoshida, J. P. Nolan, J. H. Jett and B. L. Marrone, *Anal. Chem.*, 2005, **77**, 3554.
38. H. Zipper, H. Brunner, J. Bernhagen and F. Vitzthum, *Nucleic Acids Res.*, 2004, **32**, e103.
39. J. C. Scaiano, C. Aliaga, M. N. Chretien, M. Frenette, K. S. Focsaneanu and L. Mikelsons, *Pure Appl. Chem.*, 2005, **77**, 1009.
40. K. Yagi, *Appl. Microbiol. Biotechnol.*, 2007, **73**, 1251.
41. M. Sassanfar and J. W. Roberts, *J. Mol. Biol.*, 1990, **212**, 79.
42. A. M. Chaudhury and G. R. Smith, *Mol. Gen. Genet.*, 1985, **201**, 525.
43. A. R. Fernandez De Henestrosa, T. Ogi, S. Aoyagi, D. Chafin, J. J. Hayes, H. Ohmori and R. Woodgate, *Mol. Microbiol.*, 2000, **35**, 1560.
44. R. M. Wachter, *Acc. Chem. Res.*, 2007, **40**, 120.
45. A. Norman, L. Hestbjerg-Hansen and S. J. Sorensen, *Appl. Environ. Microbiol.*, 2005, **71**, 2338.
46. A. Norman, L. H. Hansen and S. J. Sorensen, *Mutat. Res.*, 2006, **603**, 164.
47. J. H. Lee, R. J. Mitchell and M. B. Gu, *Biosens. Bioelectron.*, 2004, **20**, 475.
48. J. H. Lee, R. J. Mitchell, B. C. Kim, D. C. Cullen and M. B. Gu, *Biosens. Bioelectron.*, 2005, **21**, 500.
49. J. M. Ahn, E. T. Hwang, C. H. Young, D. L. Banu, B. C. Kim, J. H. Niazi and M. B. Gu, *Biosens. Bioelectron.*, 2009, **25**, 767.
50. J. Wang, *Anal. Chim. Acta*, 2003, **500**, 247.
51. R.-Z. Qin, C.-G. Niv, G.-M. Zeng, L. Tang and J.-L. Gong, *Talanta*, 2009, **80**, 991.
52. F. Lucarelli, I. Palchetti, G. Marrazza and M. Mascini, *Talanta*, 2002, **56**, 949.
53. F. Lucarelli, A. Kicela, I. Palchetti, G. Marrazza and M. Mascini, *Bioelectrochemistry*, 2002, **58**, 113.
54. K. R. Rogers and C. L. Gerlach, *Environ. Sci. Technol.*, 1999, **4**, 500A.
55. K. R. Rogers and M. Mascini, *Field Anal. Chem. Technol.*, 1999, **2**, 317.

CHAPTER 7
# Detection of Damage to DNA Using Electrochemical and Piezoelectric DNA-Based Biosensors

JAN LABUDA

Institute of Analytical Chemistry, Slovak University of Technology in Bratislava, 81237 Bratislava, Slovakia

## 7.1 Introduction

DNA damage is a term which denotes an alteration in chemical structure of the deoxyribonucleic acid (DNA) resulting from its interactions with physical or chemical agents occurring in the environment, generated in the organisms as by-products of metabolism, or used as therapeutics.[1] The main types of DNA damage include interruptions of the DNA sugar-phosphate backbone (strand breaks), release of the DNA bases due to hydrolysis of $N$-glycosidic bonds (resulting in abasic sites) and a variety of nucleobase lesions (adducts resulting from reactions of DNA with a broad range of genotoxic substances). Moreover, specific binding of a guest molecule by intercalation, which represents an insertion of a low molecular mass compound between the stacked base pairs of the double helix structure, can cause a lengthening of the DNA helix, perturbation of the phosphate backbone, and even untwisting of the double helix which results into accessibility of base pairs to the environment. Thus, in addition to changes of covalent bonds, the terms *DNA damage* or *subtle damage* are sometimes used to describe alterations of DNA structure induced by noncovalent binders which can affect biological function of DNA.

---

Nucleic Acid Biosensors for Environmental Pollution Monitoring
Edited by Marco Mascini and Ilaria Palchetti
© Royal Society of Chemistry 2011
Published by the Royal Society of Chemistry, www.rsc.org

Damage to DNA is feature of existence in body cells where DNA is exposed to a number of physical or chemical agents that induce damage, leading to $10^4$–$10^6$ DNA damage events per cell and per day.[2] Scientifically, damage to DNA differs from *mutation*. In damaged DNA the chemical nature of individual nucleotides is changed, whereas mutation refers to a change in DNA sequence, which can be a result of DNA damage. Serious damage to DNA is caused by chemical systems generating free radicals such as reactive oxygen (ROS), nitrogen (RNS), or sulfur (RSS) species[3,4] that oxidize the DNA bases and deoxyribose and cause release of the bases and strand breaks. Other classes of genotoxic substances include alkylating agents, substances inducing base deamination, and polycyclic aromatic hydrocarbons.[2]

Persistent DNA damage may have severe impacts on cell function and the life of the organism. The detection of DNA damage induced by environmental genotoxic agents and industrial pollutants, together with their metabolically activated products, is therefore of great importance for human health and its protection. At the same time, investigation of DNA damage by new chemotherapeutic drugs and potential DNA damage by drugs in general represent important studies in targeted development of therapeutics.[5]

For *DNA damage detection*, a wide spectrum of conventional methods such as chromatography, capillary electrophoresis, and mass spectrometry of hydrolyzed DNA samples,[6–9] gel electrophoresis and comet assay,[8,10–12] immunochemical techniques,[13,14] and others is typically used. As relatively small changes in the DNA structure (including those induced by DNA damage) significantly affect electrochemical response and behavior of DNA at voltammetric electrodes (reviewed in refs. 15–18), electroanalytical techniques are also highly sensitive tools for the detection of damage to DNA. Since the 1990s,[19] DNA, and today a wide range of of nucleic acids (NA), are utilized as biorecognition elements at a new type of biosensors denoted as DNA or NA biosensors (more precisely, DNA or NA-based biosensors) which discriminate an analyte on the bioaffinity principle. The role of physical transduction is mostly fulfilled by electrical, electrochemical, optical, and acoustic elements. The method of detection is critical in DNA biosensor or array applications. Biosensors with electrical and electrochemical transducers are popular both in the development and applications because of the general advantages of electroanalytical methods, such as rather simple sensor fabrication, low costs of equipment and analysis, possibility of miniaturization and automation, among others.[20]

Thus, an *electrochemical DNA-based biosensor* can be characterized as a device that integrates DNA (or generally a NA) as the biocomponent and an electrode as the physicochemical transducer. It is often prepared as a chemically modified electrode. The pioneering concept of an electrode modified with the DNA layer has allowed a significant decrease of the amount of DNA tested/determined.[21] A recent IUPAC technical report entitled "Electrochemical nucleic acid-based biosensors: concepts, terms and methodology" presents a critical classification of terms and techniques used in this dynamically developing field.[22]

Specific application areas of DNA-based biosensors are the detection of DNA hybridization event (so-called *genosensors*) and the detection of damage to DNA, together with investigation of association interactions of DNA relating to its damage. Electrochemical DNA biosensors have contributed significantly to basic studies of DNA damage at electrode surfaces, to the detection of low levels of genotoxic substances and their effects towards DNA, and to the evaluation of DNA protection (antioxidative) capacity of various natural and synthetic substances (reviewed in ref. 20). The electrochemical DNA-based biosensors have been used not only to detect, but also to induce and control DNA damage at the electrode surface via electrochemical generation of the damaging (usually radical) species.[1]

The aim of this chapter is to review the electrochemical and piezoelectric DNA-based biosensors and relating assays used for DNA damage detection, and a substantial part is devoted to electrochemical DNA biosensors. The chapter deals mostly with state-of-the-art detection principles and applications of DNA biosensors in recent years.

## 7.2 Electrochemical Biosensors

### 7.2.1 Construction of Biosensors for DNA Damage Detection

A chemically modified electrode[23,24] is an electrode made of a conducting or semiconducting material that is coated with selected monomolecular, multi-molecular, ionic, or polymeric film of chemical modifier, and by means of faradaic (charge transfer) reactions or interfacial potential differences exhibits chemical, electrochemical, and/or optical properties of the film. Following this principle, a thin ($<100\,\mu$m) DNA layer coverage can be deposited on the transducer. Depending on how the biosensor is fabricated, thicker layers of DNA gel often occur on the electrode surface, but this is sometimes not considered and/or reported.

Choice of the *electrode material* is connected, on one hand, with the electrochemical process of interest and, on the other hand, with DNA immobilization. The role of transducer (working electrode) is fulfilled by bulk electrodes. These are typically mercury-based (mercury film, solid amalgams), carbon-based (glassy carbon (GCE), carbon paste (CPE), graphite and pyrolytic graphite (PGE), graphite–epoxy composite), and some other (gold, platinum, indium tin-oxide (ITO)) electrodes, or by various thin- and thick-film electrodes (*e.g.*, screen-printed carbon (SPCE) and gold electrodes). DNA array sensors utilize interdigitated electrodes as the transducers.[25]

With respect to high hydrogen overvoltage, the mercury electrodes and solid amalgam electrodes (SAE) can operate at relatively high negative potentials and are suitable for studies of reduction processes. On the other hand, carbon and other solid electrodes are typically suitable for oxidation processes. Both the mercury and carbon electrodes are therefore widely used for measurement of intrinsic DNA responses. Before covering with the DNA film, a pretreatment of the bare electrode surface is sometimes performed. For instance, for CPE and sometimes for SPCE, preactivation by anodic polarization at $+1.7$ V *versus* Ag/AgCl for some time (several minutes) was suggested followed by a potential ($+0.5$ V) stimulated immobilization of dsDNA from acetate buffer solution for 120 s onto the electrode surface.[26,27]

*Nanotechnology-enabled sensors* are already widely used in the field of biosensors, including NA-based sensors (reviewed in refs. 28–30). Gold nanoparticles and carbon nanomaterials, particularly carbon nanotubes (CNT),[31–37] have attracted attention because of their unique structural, electronic, mechanical, and chemical properties. The inherent electroactivity and effective electrode surface area of CNT lead to a large enhancement of the current responses, compared to those obtained at conventional carbon electrodes.

*DNA immobilization* at the electrode surface is an initial step which plays a major role in the overall biosensor performance. Methods used for DNA film formation vary depending on the kind of transducer and the biosensor application, and detailed experimental conditions have to be optimized for each special application.[20,38] Numerous surface and "bulk" phase DNA modified electrodes have been reported.[39] The methods range from noncovalent DNA binding by physisorption (reviewed in refs. 18,20), entrapment within polymeric films (reviewed in refs. 40,41), affinity binding (*e.g.*, by the extremely strong avidin–biotin system[42]), chemisorption (by self-assembled monolayer formation[43,44]), formation of layer-by-layer assembly,[45] to direct covalent binding (*e.g.*, by the carbodiimide method[46] or covalent attachment to the synthetic polymer films bearing reactive linkers such as biotin, complexation ligand, *etc.*).

Atomic force microscopy (magnetic A/C mode AFM) images in air were used to characterize two different procedures for immobilizing nanoscale double-stranded (ds) DNA surface films on carbon electrodes. Thin-film dsDNA layers exhibited holes in the dsDNA film that left parts of the electrode surface uncovered, while thicker films showed a uniform and complete coverage of the electrode.[47,48]

## 7.2.2 Techniques Used for DNA Damage Detection

Electrochemical measurements with DNA biosensors are mostly performed in voltammetric and chronopotentiometric detection modes.[20] With the general progress in impedimetric biosensors, electrochemical impedance spectroscopy (EIS) has become popular as the measurement technique for electrical DNA-based biosensors.[49,50] According to the electrochemically active species for which responses are evaluated at in the detection of DNA damage, the experimental techniques can be classified as follows:[22]

- Label-free techniques which use no chemical modification of DNA or substances interacting with it by specific labels.
- Techniques which employ electrochemically active labels covalently bound to DNA.
- Reagentless techniques, in which no additional chemical reagents (indicators, redox mediators, enzyme substrates) are needed to generate an analytical signal.
- Techniques that employ redox indicators either noncovalently bound to the DNA (groove binders, intercalators, anionic or cationic species interacting with DNA electrostatically) or present in the solution phase.

### 7.2.2.1 Label-free and Indicatorless Detection Techniques

These techniques utilize electrochemical (charge transfer) activity and/or surface activity of DNA itself at voltammetric electrodes.[18] The electrochemical activity of DNA is based on the presence of redox active sites at nucleobases and sugar residues. Adenine, cytosine, and guanine give electrochemical reduction responses at mercury-based electrodes in neutral and weakly acidic media, and the guanine residues also yield an anodic signal due to electrooxidation of the reduction product back to guanine. All common nucleobases can be electrochemically oxidize at carbon and some other solid electrodes such as gold, platinum, and indium tin oxide (ITO) electrodes. While mercury and mercury-amalgam electrodes are excellent transducers for the label-free detection of DNA strand breaks (and, in connection with DNA repair enzymes, the detection of some types of nucleobase lesions), the intrinsic DNA responses at carbon electrodes are inherently less sensitive to changes in the DNA structure upon small levels of DNA damage. The reasons are relatively easy accessibility of the guanine moiety to electrochemical oxidation via major groove of the helix, and the absence of extensive DNA denaturation at the surface of carbon electrodes.[17] As the electrochemical reduction and oxidation of DNA bases are both irreversible, measurements with the biosensors cannot be performed repeatedly.

The polyanionic nature of the NAs leads to characteristic adsorption/desorption (reorientation) processes at the mercury-based electrodes upon applying negative electrode potential, due to an interplay between electrostatic repulsion and relatively strong DNA adsorption via hydrophobic parts of the polynucleotide chains, particularly the DNA bases.[1,18] The DNA response in weakly alkaline background electrolytes is sensitive to minor conformational changes of DNA induced by nucleases and various chemical as well as physical agents including ionizing radiation and ultrasound.[51]

*Detection of Strand Breaks with Mercury-Based Sensors.* Strand breaks (SB) are frequent DNA lesions due to acidic or enzymatic hydrolysis of the phosphodiester bonds, damage to the deoxyribose moiety (*e.g.*, by radical species) and/or as a result of some kinds of DNA base damage. At mercury-based electrodes, electrochemical behavior of DNA strongly depends on its backbone structure which influences accessibility of the DNA bases residues for the electrode reaction. In single-stranded (SS) DNA, the bases can easily exchange electrons with the electrode and possess specific responses (the reduction of cytosine and adenine, and the tensammetric peak). On the other hand, in dsDNA, the bases are situated in the interior of the double helix and the native DNA is electrochemically inactive. However, around the ends of linear (lin) DNA molecules and around single-strand breaks (SSB), the double helix undergoes transient opening which results in specific signals. DNA damage by various DNA-breaking agents was detected using DNA-modified electrodes and these principles.[17,51]

Using a slow scan from positive to negative potential values, surface denaturation of dsDNA is induced (reviewed in refs. 16,18), which also leads to the formation of a DNA chain with free ends possessing the responses specific for ssDNA (which are not produced by intact dsDNA). In the case of closed circular DNA, such a denaturation is impossible for topological reasons.

The effects of various DNA cleavage agents were also observed. Potential-dependent cleavage of supercoiled (sc) DNA by ROS formed in the Fenton reaction (generated by iron or copper complex compounds and hydrogen peroxide or oxygen reduced electrochemically to hydrogen peroxide) can serve as an example.[52] By using alternating current (AC) voltammetry, one SB was detected among more than $2\times10^5$ nucleotides.[53] With the detection limit at the femtomole level in one measurement, this approach is comparable with conventional DNA damage assays.[54]

Although a conventional hanging mercury drop electrode (HMDE) used as the transducer possesses such unique features, successful attempts have been made to replace it by other electrodes in which the liquid mercury content is minimized or eliminated. Both redox and tensammetric DNA responses were obtained at a mercury film-coated glassy carbon electrode (MF/GCE) and different variants of the silver solid amalgam electrodes (AgSAE). Responses of sc, lin, and ssDNA at AgSAE and MF-AgSAE electrodes are analogous to those observed at HMDE.[55,56] MF/GCE[57] as well as AgSAE and MF-AgSAE[55] electrodes modified with scDNA were used to detect nicking of plasmid DNA with enzymes such as DNase I and reactive radical species which destroy the deoxyribose moieties. Some types of nucleobase lesions after their conversion to strand breaks by specific enzymes, as well as a reverse process (*i.e.,* the repair of strand breaks by action of the DNA ligases), were detected using the same scheme.

*Biosensors Based on Guanine Redox Processes.* Procedures based on intrinsic responses due to cathodic reduction/anodic reoxidation and anodic oxidation of the guanine residues are among the most widely used of the electrochemical DNA biosensors because (1) guanine has well-defined responses at both mercury- and carbon-based electrodes, and (2) guanine is, among the DNA bases, the most frequent target for a broad range of genotoxic agents (reviewed in refs. 58, 59). Chemically or photochemically induced chemical changes in the DNA guanine moiety may alter its electrochemical behavior and response corresponding to loss of the parent base. Such changes were studied by using DNA-modified HMDE.[60–62] At graphite electrodes, intrinsic voltammetric responses of dsDNA (due to oxidation of guanine and adenine residues) were increased as a result of exposing the electrode to sufficiently negative potentials (between –0.4 and – 0.8 V) prior to potential scanning.[63] It was concluded that dsDNA was unwound at the negatively charged graphite surface, similarly to what was observed at mercury electrodes.

Various DNA-modified carbon electrodes such as GCE, pyrolytic graphite electrode (PGE), CPE, and disposable screen-printed carbon electrode (SPCE) were also utilized and effects of genotoxic agents such as hydrazines,[64] platinum compounds,[65] metal complexes,[66] various aromatic compounds,[67,68] arsenic compounds,[69,70] and UV light[71] were detected.

Decrease of the anodic guanine peak height or area relative to that yielded by intact DNA was suggested as a measure representing damage to this nucleobase. A disposable electrochemical DNA biosensor based on dsDNA immobilized on the surface of a graphite screen-printed electrode (SPE) was proposed as a screening test for environmental toxicants present in water or wastewater samples.[72-74] DNA modification was estimated with the portion of response decrease ($R$, %) which is the ratio of the guanine peak area after the interaction with the analyte (GPAs), and the guanine peak area after the interaction with the buffer solution (GPAb): $R$, % = [(GPAs/GPAb)–1]×100.

The result of the test for one sample can be obtained within 8 min.[75] Comparison of the results obtained with the DNA biosensor with those obtained by using a commercial luminescent bacteria test, Toxalert 100, showed some difference between the two methods. While the DNA biosensor response indicates mainly the interaction of DNA with some chemicals which decrease the guanine oxidation capability, the Toxalert response is more complex since the metabolic activity of the bacterial cell is involved. Nevertheless, the results of the real sample analysis showed a promising correlation between the two tests.

Wastewater samples provided during First European Interlaboratory Exercise on water toxicity in the course of the project SWIFT-WFD were analyzed, and biosensor results were compared with a currently used toxicity test Tox-Alert 100 based on the bioluminescence inhibition of *Vibrio fischeri*. The results showed a promising correlation between two tests used for the detection of toxic compounds in water samples.[76] Also other toxicity tests, such as the comet test, show a similar trend when a standard solution as well as real samples are analysed, thus confirming the powerful application potential of these rapid and inexpensive devices. Another similar investigation of potential chemotherapeutics was reported.[77]

As natural DNA contains many guanine residues, only a partial decrease of the guanine peak is usually observed, depending on the extent of DNA damage. Decrease in the guanine response is obviously also caused by a release of the base from the polynucleotide chains, an event which often follows modifications within the guanine imidazole ring, and/or by a release of DNA from the electrode surface as a result of deep SBs. Such detection of DNA damage is sometimes more sensitive to the effect and concentration of DNA-damaging species than the use of other techniques. Nevertheless, the guanine anodic peak was also reported first to increase and then to decrease in height, indicating a more complex DNA damage profile, for instance helix opening followed by its degradation.[78]

As only guanine moieties in the close vicinity of the electrode surface can undergo direct electro-oxidation, soluble *redox mediators* such as rhodium or ruthenium complexes are sometimes used to shuttle electrons from guanine residues in distant parts of DNA chains to the electrode;[79-81] however, this can no longer be considered a reagentless technique. For instance, natural dsDNA immobilized on the electrode surface by layer-by-layer assembly within poly (styrene sulfonate)/poly(diallyldimethylammonium)/dsDNA (PSS/PDDA/dsDNA) films was used to detect DNA lesions induced by the alkylating agent methyl methanesulfonate (MMS) by cyclic voltammetry with the ruthenium(II) tris(2,2′-bipyridyl) complex, $[(Ru(bpy)_3]^{2+}$, in solution.[82] After treatment by the *E. coli* exonuclease III enzyme, the electrocatalytic oxidation peak of the films was further amplified and greatly enhanced because the enzyme could convert those apurinic sites caused by MMS in the damaged dsDNA into ssDNA regions, so that more guanines in the DNA became exposed.

### 7.2.2.2  Biosensors with Labeled DNA

Electroactive labels introduced into DNA allow us to obtain electrochemical signals at less extreme potentials than those typical for intrinsic DNA responses. The first electroactive labels covalently bound to DNA were electroactive osmium labels, introduced in the early 1980s. Osmium tetroxide complexes ($Os^{VIII}$,L) with nitrogen ligands like pyridine and bipyridine bind to pyrimidine residues in ssDNA and are known in connection with DNA sequencing techniques. Another label is ferrocene, perhaps the most frequently used electroactive label in DNA hybridization sensors. Ferrocene and some other labels require solid-state organic chemistry, and unlike the Os,L labels they could hardly be used for labeling of longer NA, such as plasmid or chromosomal DNA. Other compounds, such as daunomycin, viologen, and thionine have been used as electroactive labels of ODNs. Electroactive labels can be introduced into DNA not only by chemical methods, but also by means of enzymes such as DNA polymerases (reviewed in ref. 18).

An approach to the electrochemical sensing of DNA damage using carbon electrodes and osmium tetroxide complexes has been described recently.[83] The technique is based on enzymatic digestion of DNA with a DNA repair enzyme exonuclease III (exoIII), followed by chemical modification of DNA with a complex of osmium tetroxide with 2,2′-bipyridine bound to free 3′-ends of single-stranded regions created by the exoIII. Intensity of the electroactive osmium label response is measured at a pyrolytic graphite electrode (PGE) and responds to the extent of DNA damage. The technique was used for the detection of single-strand breaks (SSB) introduced in plasmid DNA by deoxyribonuclease I and apurinic sites generated in chromosomal calf thymus DNA upon treatment with the alkylating agent dimethyl sulfate. The apurinic sites were converted into SSB by AP-nuclease activity of the exoIII enzyme. The technique is capable of detection of one lesion per approximately $10^5$ nucleotides in plasmid scDNA.

Recently, DNA hybridization biosensors for studies of DNA damage by common toxicants and pollutants have been proposed.[84] Voltammetric transduction was achieved by coupling a ferrocene moiety to streptavidin linked to biotinylated target DNA. The sensor responds rapidly to any damage caused by Cr(VI) species, with more severe DNA damage observed for $Cr_2O_7^{2-}$ and for $CrO_4^{2-}$ in the presence of $H_2O_2$ as compared to $CrO_4^{2-}$ alone. The herbicides and pesticides examined caused DNA damage or structural alterations leading to the double-helix unwinding. Among these compounds, paraoxon-ethyl and atrazine caused the fastest and most severe damage to DNA. Physicochemical mechanisms have been proposed for the damaging interactions between toxicants and DNA, as well as the kinetics of DNA damage and unwinding.

### 7.2.2.3 Redox Indicator-Based Response to DNA Damage

A number of detection techniques with electrochemical DNA biosensors employ electroactive species added to the system and *interacting noncovalently with DNA* as its indicators. These are label-free methods, as DNA is not chemically modified by a special label; however, they are not reagentless techniques. Like DNA labels, redox indicators typically have electrochemical responses at a "safe" electrode potential, and often reversibly. (The terms "redox probe" and "redox marker" are sometimes found in the literature, but this usage could cause confusion with the DNA capture probe used as a recognition element in DNA hybridization and with markers used in medical diagnostics.[22])

Some of the redox indicators interact with DNA on the basis of electrostatic forces.[85] Cationic indicators such as metal complex cations are attracted to the DNA by the negative charge of the DNA backbone. For instance, the cobalt complex with 1,10-phenanthroline, $[Co(phen)_3]^{3+}$, was repeatedly used to detect DNA degradation in CPE- and SPCE-based biosensors by chemical systems including those producing ROS.[34,68,86,87] Other DNA redox indicators intercalate into the dsDNA structure (*e.g.*, daunomycin[88], methylene blue[89,90]) or bind to dsDNA groove (*e.g.*, bis-benzimide—a fluorescent dye, Hoechst 33258). All cationic indicators, intercalators, and groove binders accumulate at the immobilized dsDNA layer (prior to damage), and thus give rise to increased voltammetric response.

Using the scheme of indicator accumulation/voltammetric measurement/ chemical regeneration of the DNA layer (by removal/desorption of the accumulated indicator particles), the response of the indicator can be measured repeatedly. Then, using one single biosensor, a mean indicator response value prior to DNA damage and its standard deviation are calculated. Damage to DNA (*e.g.*, deep degradation including SB) during an incubation of the biosensor in the cleavage agent for a given time (minutes to hours) followed by medium exchange and repeated voltammetric measurement results in diminution of the original indicator voltammetric response (signal-off technique) depending on the degree of DNA damage. Thus, the proportion of DNA that has survived incubation of the biosensor in the cleavage agent can be estimated.[34,68,69]

Sensitive electrochemical sensing for DNA damage *in situ* based on a cathodic process with a $Fe@Fe_2O_3$ core–shell nanonecklace and gold nanoparticles was performed by a novel biosensor, which was constructed via a GCE modified with a multilayer film made up of separate layers of PDDA, the mixture of $Fe@Fe_2O_3$ core–shell nanonecklace and gold nanoparticles, PDDA, and dsDNA.[91] Iron ions and $H_2O_2$ (Fenton reagents) were generated continuously at a constant rate by the cathodic process. The two Fenton reagents reacted further to generate hydroxyl radicals *in situ*, which attacked dsDNA in the film and caused severe damage to it. The biosensor was shown to be a potential screening tool for rapid assessment of the genotoxicity of existing and new chemicals. In a continuation of this work, another electrochemical biosensor for mimicking the metal-mediated DNA damage pathway *in situ* was presented.[92] It is based on a Fe@Fe$_2$O$_3$ core-shell nanonecklace and multiwalled CNT composite. The Fenton reagents, H$_2$O$_2$ and iron ion, react together to yield a hydroxyl radical, which attacked DNA in the film. The DNA damage was detected by monitoring the differential pulse voltammetric response of an electrochemical indicator, Co(phen)$_3^{3+}$. Another electrochemical indicator, Ru(NH$_3$)$_6^{3+}$, was also used as a complementary means of monitoring the DNA damage, and the minimal detectable amount of DNA damage was 0.16 μg.

An electrochemical sensor for detecting damage of natural dsDNA was constructed forming PSS/{PDDA/dsDNA}$_3$ layer-by-layer films assembled on pyrolytic graphite (PG) electrodes.[93] When the PSS/{PDDA/dsDNA}$_3$ films were immersed into MB solution and MB loaded into the films, the PSS/{PDDA/dsDNA}$_3$-MB films formed in blank buffers at pH 7.0 showed a reversible cyclic voltammetric peak pair of MB at –0.23 V *versus* SCE and good reversibility of MB incorporation. After incubation in a solution of the known genotoxic agent styrene oxide (SO), the damaged PSS/{PDDA/dsDNA}$_3$-MB films could not return to their original, fully loaded state with reloading of MB, and showed smaller cyclic voltammetry (CV) peak currents than those of intact PSS/{PDDA/dsDNA}$_3$-MB films. The relative peak current ratio, $I_{p,I}/I_{p,III}$, where $I_{p,I}$ was the anodic peak current of intact DNA films in blank buffers after fully loading with MB and $I_{p,III}$ was that of SO-damaged DNA films after reloading with MB, increased linearly with the incubation time with SO in the range of 5–40 min with a damage rate of 0.0099 min$^{-1}$. Steric hindrance of SO adducts with guanine or adenine blocked the intercalation of MB into the base pairs of dsDNA, resulting in the decrease of CV peak currents of loaded MB. The specific intercalation of MB into dsDNA base pairs and the sensitive electrochemical response of MB, combined with the unique feature of loading reversibility of MB in the DNA layer-by-layer films, makes the difference in CV response between the intact and damaged dsDNA films become pronounced in the "loading/release/reloading" procedure.

Electrochemical indicators binding preferentially to either ss or dsDNA have been utilized to detect DNA hybridization as well as DNA damage, for instance for the recognition of intact dsDNA from degraded DNA that has lost its double-helical structure.[1,20]

The redox indicators may be also used as *diffusionally free species* present as reagents in the solution phase. For instance, hexacyanoferrate(III/II) anions, [Fe(CN)$_6$]$^{3-/4-}$, are used in CV experiments where they indicate the presence of DNA layer on the electrode surface on the basis of electrostatic repulsion between the indicator anion and negatively charged DNA backbone, which leads to a more poorly developed CV picture with decreased peak current and increased anodic to cathodic peak potential separation.[37,68,87]

The anionic complex [Fe(CN)$_6$]$^{3-/4-}$ serves also as typical redox indicator widely utilized for the characterization of electrode surfaces and chemically modified electrodes by EIS. It has also been used in DNA impedimetric biosensors for the detection of DNA damage as well as its protection.[33,34,37,88]

Both cationic and anionic redox indicators have also been successfully used for the evaluation of antioxidative behavior and strength of natural and synthetic compounds[34,94–96] as well as tea and plant extracts.[33,97–99] Here, the experimental procedure is based on incubation of the DNA biosensor in a mixture of cleavage agent and antioxidant, followed by the evaluation of the indicator response as described above. Experiments were also performed in a flow-through electrochemical system.[100]

The interaction of •OH radicals, generated via a Fenton-type reaction, with *immobilized DNA* in the absence and presence of antioxidants was evaluated by means of changes in the guanine oxidation peak obtained by square wave voltammetry (SWV). The results demonstrated that the DNA-based biosensor is suitable as a rapid *in vitro* screening test for the evaluation of antioxidant properties of plant extracts.[101] Recently, electrochemical biosensors developed for the evaluation of the antioxidant capacity of specific compounds were critically reviewed. Three different sensing approaches are described, based on cytochrome *c*, superoxide dismutase, and DNA. Because of the ability of these devices to perform simple, fast, and reliable analysis, they are promising biotools for the assessment of antioxidant properties.[102]

## 7.2.2.4 Electroactive Products of DNA Damage

Some of the products of DNA damage exhibit characteristic electrochemical activity. New electrochemical signals can be evaluated with better sensitivity (so-called *signal-on technique*) than the change in the original guanine response (*signal-off technique*). Typically, the DNA biosensor is incubated at a constant negative potential for a period of time in order to preconcentrate short-living radical intermediates formed during the electrochemical reduction of the analyte under investigation. These radicals, when in very close contact with DNA, can rapidly interact causing DNA damage which is detected on carbon electrodes by the occurrence of anodic peak corresponding to the DNA purine base (guanine moiety at 0.9 V *versus* SCE) and another peak at +0.4 V *versus* SCE attributed to the anodic oxidation of 8-oxoguanine.[103] Thus, 8-OG is generated by the interaction between the radicals and a neighboring guanine in the dsDNA during the cathodic polarization of the biosensor (reviewed in refs. 1, 20, 59). Besides the direct detection of 8-OG in the presence of guanine, an Os(III/II) mediated electrochemical oxidation of 8-OG was reported.[104,105]

Some drugs and chemicals such as niclosamide,[106] adriamycin,[107–109] copper complex with quercetin,[109–111] benznidazole,[112] thiophene-*S*-oxide,[113] and nitroderivatives of polycyclic aromatic compounds were investigated in this way.[114]

Formation of 2,8-dihydroxyadenine, the oxidation product of adenine residues and a biomarker of DNA oxidative damage, was also detected using a DNA biosensor and differential pulse voltammetric detection mode for *in situ* evaluation of the interaction of heavy metals with dsDNA.[115] It was found that

$Pb^{2+}$, $Cd^{2+}$, and $Ni^{2+}$ bind to dsDNA, and that this interaction leads to different modifications in the dsDNA structure which can be electrochemically recognized as changes in the oxidation peaks of guanosine and adenosine. Using homopolynucleotides of guanine and adenine it was proved that the interaction between $Pb^{2+}$ and DNA causes oxidative damage and preferentially takes place at adenine-containing segments, with the formation of 2,8-dihydroxyadenine. The interaction of $Cd^{2+}$ and $Ni^{2+}$ causes conformational changes, destabilizing the double helix, which can enable the action of other oxidative agents on DNA.

The damaged DNA modified by "bulky" adducts to the nucleobases (e.g., mitomycin C and other drugs whose pharmacological effects involve DNA modification) is also able to give specific electrochemical response of the modifying moieties.[116]

### 7.2.2.5 Layered Assemblies for Genotoxicity Screening

Biosensors based on nanometer-scale complex films of DNA, enzymes, polyions, and catalytic redox-active ruthenium polyions were suggested for tests of genotoxic activity of various chemicals (reviewed in ref. 117). Here, enzymatically active hemoproteins mimic the metabolic activation processes of a carcinogen. For instance, styrene was enzymatically converted by heme proteins to SO, which diffused into the DNA layer where it attacked the guanine moiety. The consequent opening of the DNA double helix facilitated electrocatalytic oxidation of other guanine moieties mediated by the ruthenium complex.[118,119]

Simultaneous optical and voltammetric detection of bioactivated genotoxicity was also reported employing ultrathin films of DNA, model metabolic enzymes (cytochrome $P450_{cam}$ and myoglobin), and electrochemiluminescence (ECL) generating metallopolymer $[Ru(bpy)_2PVP_{10}]^{2+}$ on pyrolytic graphite (PG) electrodes.[120] Sensor film growth and component amounts were monitored using a quartz crystal microbalance (QCM). DNA damage in the sensor films was evaluated simultaneously using a simple apparatus combining a standard voltammetry cell coupled with an optical fiber and photomultiplier tube. The model enzyme reaction converted styrene to SO, which reacts with DNA nucleobases. Electrochemiluminescent and SWV signals increased with enzyme reaction time on the scale of several minutes, and provided relative enzyme turnover rates for DNA damage suitable for toxicity screening applications. Within 1 min, the sensor detects about 3 damaged bases per 10 000 DNA bases using this simultaneous detection.

### 7.2.2.6 Molecular Beacon-like Sensor for Nuclease and Ligase Activities

An electrochemical biosensor based on a hairpin DNA probe labeled with ferrocene was reported for monitoring the activities of nucleases (generating single-strand breaks) and DNA ligases (sealing the strand break).[121] The stem (duplex) part of the hairpin structure contained a single strand break and the ferrocene-labeled segment was removed under denaturing conditions. In the presence of ligase activity, the break was joined, preventing removal of the ferrocene-labeled segment and resulting in the appearance of a current signal due to the ferrocene oxidation. When the continuous form of the hairpin (without the break) was exposed to a restriction nuclease, the same procedure resulted in diminution of the current signal.

## 7.3 Piezoelectric Biosensors

A piezoelectric (mass-sensitive) quartz crystal sensor consists of a thin quartz disc sandwiched between a pair of electrodes. Quartz is a piezoelectric material that deforms when an electric field is applied across the electrode. The resonance frequency of the quartz crystal depends on the total oscillating mass. Thus, piezoelectric materials offer an attractive, near-universal mode of transducing any biorecognition event, if the changes in detector mass that accompany analyte binding are sufficiently large. Resolution of mass changes of less than $1 \times 10^{-9}$ g cm$^{-2}$ is possible in liquid media, at least for substances of high molecular mass.[122] Hence, the sensor is capable of measuring very small mass changes.[123] Piezoelectric transducers also offer the advantages of solid-state construction, chemical inertness, durability, the possibility of low-cost mass production, and rapid real-time detection.

A piezoelectric quartz crystal sensor is a thin DNA film deposited on a sensitive piezoelectric quartz crystal. This noninvasive gravimetric technique is based on a measurement of change in the resonance frequency resulting from the change of mass of the DNA film. Increase in film mass accompanies a hybridization event,[124] and decreased mass is connected with a DNA degradation. A sensor for real-time probing of dynamic enzymatic DNA cleavage processes was prepared by thermal evaporation of 100-nm gold onto a predeposited chromium underlayer on a quartz matrix, followed by immobilization of thiolated ODN by the self-assembly technique. Mass changes associated with the enzymatic digestion indicated activity and specificity of nucleases.[125] Oxidative damage to DNA by the $H_2O_2$ system containing $Cu^{2+}$ or $Zn^{2+}$ was studied.[126]

Piezoelectric quartz crystal impedance (PQCI) analysis is another kind of piezoelectric sensing technique. It can provide not only the resonant frequency ($f_0$), but also other parameters of the quartz crystal sensor, such as motional resistance ($R_m$), motional inductance ($L_m$), motional capacitance ($C_m$), and static capacitance ($C_0$). Therefore, PQCI can provide multidimensional information about the process that has occurred at a solid–liquid interface. Investigation of DNA damage by quercetin–$Cu^{2+}$ ions was reported where potentiomentric stripping signals of guanine and adenine were monitored as well.[127]

In quartz crystal microbalance with dissipation monitoring (QCM-D), the change in resonance frequency relates to the adsorbed mass, and the change in dissipation is a measure of the dampening effect of the adsorbed film and provides an indication of the rigidity of the film. This system was used for detection of enzymatic cleavage of surface-tethered DNA oligomers.[128,129]

Monitoring mammalian DNA damage by using DNA adsorbed to a polyelectrolyte surface and exposed to quercetin–Cu(II) agent inducing DNA strand scission was also reported.[130]

The electrochemical quartz crystal nanobalance (EQCN) has been used recently to detect the DNA hybridization process, and prepared nanogravimetric and voltammetric DNA biosensors have been applied to studies of DNA damage by common toxicants and pollutants.[131]

## 7.4 Conclusions

The role of DNA as one of the most important biomacromolecules and investigation of its (biochemical/chemical) reactivity has increased the need for simple, fast, cheap, miniaturized, and mass-producible analytical devices which fulfill some basic criteria for decentralized DNA testing and make it possible to enter the growing market for molecular diagnostics. Similarly to other branches of biochemical analysis applied to clinical and environmental tests, such as detection of glucose and pollutants, where biosensors are widely used, the investigation of DNA sequencing and damage has led, and continues to lead, to great progress in the development and application of DNA-based biosensors.[25] Considering the high chemical stability of DNA relative to other recognition elements such as enzymes, antibodies, or other selective proteins, the analytical requirements for the construction of disposable and easy-to-use DNA biosensors can be fulfilled.

In this chapter, construction and detection principles for DNA damage biosensors (mainly electrochemical types) have been presented systematically, together with their main application areas. The DNA-based biosensors were shown to be devices that make it possible to evaluate and maybe also to classify the mode of genotoxic effects of known and unknown compounds and complex samples. Signal transduction and detection modes, or a combination of the detection modes, still plays a key role in both sensitivity and selectivity of DNA damage investigation. Before long the time will come when these devices will be used in prescreening of new drugs and newly synthesized chemicals, as well as food and environmental contaminants.

## References

1. M. Fojta, in *Electrochemistry of Nucleic Acids And Proteins. Towards Electrochemical Sensors for Genomics and Proteomics*, ed. E. Palecek, F. Scheller and J. Wang, Elsevier, Amsterdam, 2005, pp. 386–431.
2. E. C. Friedberg, *Nature*, 2003, **421**, 436.
3. M. S. Cooke, M. D. Evans, M. Dizdaroglu and J. Lunec, *FASEB J.*, 2003, **17**, 1195.
4. A. Barzilai and K.-I. Yamamoto, *DNA Repair*, 2004, **3**, 1109.
5. N. Saijo, T. Tamura and K. Nishio, *Cancer Chemoth. Pharm.*, 2003, **52**, S97.

6. A. Jenner, T. G. England, O. I. Aruoma and B. Halliwell, *Biochem. J.*, 1998, **331**, 365.
7. S. Inagaki, Y. Esaka, M. Sako and M. Goto, *Electrophoresis*, 2001, **22**, 3408.
8. A. Collins, C. Gedik, N. Vaughan, S. Wood and A. White *et al.*, *Free Radical Biol. Med.*, 2003, **34**, 1089.
9. M. Birincioglu, P. Jaruga, G. Chowdhury, H. Rodriguez, M. Dizdaroglu and K. S. Gates, *J. Am. Chem. Soc.*, 2003, **125**, 11607.
10. M. Fojta and E. Palecek, *Anal. Chim. Acta*, 1997, **342**, 1.
11. N. Yamashita, H. Tanemura and S. Kawanishi, *Mutat. Res.*, 1999, **425**, 107.
12. J. P. Pouget, T. Douki, M. J. Richard and J. Cadet, *Chem. Res. Toxicol.*, 2000, **13**, 541.
13. J. Cadet, C. D'Ham, T. Douki, J. P. Pouget, J. L. Ravanat and S. Sauvaigo, *Free Radical Res.*, 1998, **29**, 541.
14. G. Guetens, G. De Boeck, M. Highley, A. T. van Oosterom and E. A. de Bruijn, *Crit. Rev. Clin. Lab. Sci.*, 2002, **39**, 331.
15. E. Palecek and M. Fojta, *Anal. Chem.*, 2001, **73**, 74A.
16. E. Palecek, M. Fojta, F. Jelen and V. Vetterl, in *The Encyclopedia of Electrochemistry*, Vol. 9 *Bioelectrochemistry*, ed. A. J. Bard and M. Stratsmann, Wiley-VCH, Weinheim, 2002, p. 365.
17. M. Fojta, *Electroanalysis*, 2002, **14**, 1449.
18. E. Palecek and F. Jelen, in *Electrochemistry of Nucleic Acids and Proteins. Towards Electrochemical Sensors for Genomics and Proteomics*, ed. E. Palecek, F. Scheller and J. Wang, Elsevier, Amsterdam, 2005, pp. 74–174.
19. K. M. Millan and S. R. Mikkelsen, *Anal. Chem.*, 1993, **65**, 2317.
20. J. Labuda, M. Fojta, F. Jelen and E. Palecek, in *Encyclopedia of Sensors*, ed. C. A. Grimes, E. C. Dickey and M. V. Pishko, American Scientific Publishers, Stevenson Ranch, CA, 2006, pp. 201–228.
21. E. Palecek and I. Postbieglova, *J. Electroanal. Chem.*, 1986, **214**, 359.
22. J. Labuda, A. M. O. Brett, G. Evtugyn, M. Fojta, M. Mascini, M. Ozsoz, I. Palchetti, E. Paleček and J. Wang, *Pure Appl. Chem.*, 2010, **82**, 1161.
23. R. A. Durst, A. J. Baumner, R. W. Murray, R. P. Buck and C. P. Andrieux, *Pure Appl. Chem.*, 1997, **69**, 1317.
24. W. Kutner, J. Wang, M. L'Her and R. P. Buck, *Pure Appl. Chem.*, 1998, **70**, 1301.
25. A. Sassolas, B. D. Leca-Bouvier and L. J. Blum, *Chem. Rev.*, 2008, **108**, 109.
26. J. Wang, X. H. Cai, B. M. Tian and H. Shiraishi, *Analyst*, 1996, **121**, 965.
27. F. Lucarelli, I. Palchetti, G. Marrazza and M. Mascini, *Talanta*, 2002, **56**, 949.
28. J. Wang, in *Electrochemistry of Nucleic Acids and Proteins. Towards Electrochemical Sensors for Genomics and Proteomics*, ed. E. Palecek, F. Scheller and J. Wang, Elsevier, Amsterdam, 2005, pp. 369–384.
29. A. M. O. Brett, in *Electrochemistry at the Nanoscale*, ed. P. Schmuki and S. Virtanen, Springer, New York, 2009, pp. 407–433.

30. A. Ferancova and J. Labuda, in *Nanostructured Materials in Electrochemistry*, ed. A. Eftekhari, Wiley-VCH, Weinheim, 2008, pp. 409–434.
31. A. Ferancová, R. Ovádeková, M. Vaníčková, A. Šatka, R. Viglaský, J. Zima, J. Barek and J. Labuda, *Electroanalysis*, 2006, **18**, 163.
32. R. Ovádeková, S. Jantová, S. Letašiová and J. Labuda, *Anal. Bioanal. Chem.*, 2006, **386**, 2055.
33. G. Ziyatdinova, J. Galandová and J. Labuda, *Int. J. Electrochem. Sci.*, 2008, **3**, 223.
34. J. Galandová, G. Ziyatdinova and J. Labuda, *Anal. Sci.*, 2008, **24**, 711.
35. S. Flickyngerova, R. Ovadekova, I. Novotny, V. Tvarozek, J. Labuda, V. Breternitz and C. Knedlik, *Vacuum*, 2008, **82**, 303.
36. J. Galandova, L. Trnkova, R. Mikelova and J. Labuda, *Electroanalysis*, 2009, **21**, 563.
37. J. Galandová, R. Ovádeková, A. Ferancová and J. Labuda, *Anal. Bioanal. Chem.*, 2009, **394**, 855.
38. J. Wang, in *Electrochemistry of Nucleic Acids and Proteins. Towards Electrochemical Sensors for Genomics and Proteomics*, ed. E. Palecek, F. Scheller and J. Wang, Elsevier, Amsterdam, 2005, pp. 175–194.
39. M. Vanickova, M. Buckova and J. Labuda, *Chem. Anal.*, 2000, **45**, 125.
40. P. Mailley, A. Roget and T. Livache, in *Electrochemistry of Nucleic Acids and Proteins. Towards Electrochemical Sensors for Genomics and Proteomics*, ed. E. Palecek, F. Scheller and J. Wang, Elsevier, Amsterdam, 2005, pp. 297–329.
41. J. Galandova and J. Labuda, *Chem. Pap.*, 2009, **63**, 1.
42. H. Chen, C. K. Heng, P. D. Puiu, X. D. Zhou, A. C. Lee, T. M. Lim and S. N. Tan, *Anal. Chim. Acta*, 2005, **554**, 52.
43. J. Watterson, P. A. E. Piunno and U. J. Krull, *Anal. Chim. Acta*, 2002, **469**, 115.
44. J. J. Gooding, F. Mearns, W. R. Yang and J. Q. Liu, *Electroanalysis*, 2003, **15**, 81.
45. Y. Zhang, H. Zhang and N. Hu, *Biosens. Bioelectron.*, 2008, **23**, 1077.
46. K. M. Millan, A. J. Spurmanis and S. R. Mikkelsen, *Electroanalysis*, 1992, **4**, 929.
47. A.-M. Chiorcea and A. M. Oliveira Brett, *Bioelectrochemistry*, 2004, **63**, 229.
48. A. M. Oliveira-Brett, A. M. Chiorcea Paquim, V. C. Diculescu and J. A. P. Piedade, *Med. Eng. Phys.*, 2006, **28**, 963.
49. E. Katz and I. Willner, in *Technology and Performance*, ed.V. Mirsky, Springer-Verlag, Berlin, 2004, pp. 67–106.
50. J.-Y. Park and S.-M. Park, *Sensors*, 2009, **9**, 1.
51. M. Fojta, *Collect. Czech Chem. Commun.*, 2004, **69**, 715.
52. M. Fojta, L. Havran, T. Kubicarova and E. Palecek, *Bioelectrochemistry*, 2002, **55**, 25.
53. M. Fojta and E. Palecek, *Anal. Chim. Acta*, 1997, **342**, 1.
54. K. Cahova-Kucharikova, M. Fojta, T. Mozga and E. Palecek, *Anal. Chem.*, 2005, **77**, 2920.

55. R. Fadrna, K. Kucharikova-Cahova, L. Havran, B. Yosypchuk and M. Fojta, *Electroanalysis*, 2005, **17**, 452.
56. K. Kucharikova, L. Novotny, B. Yosypchuk and M. Fojta, *Electroanalysis*, 2004, **16**, 410.
57. T. Kubicarova, M. Fojta, J. Vidic, L. Havran and E. Palecek, *Electroanalysis*, 2000, **12**, 1422.
58. E. Palecek and M. Fojta, in *Electrochemical DNA Sensors, Bioelectronics*, ed. I. Wilner and E. Katz, Wiley-VCH, Weinheim, 2005, pp. 127–192.
59. A. M. O. Brett, V. C. Diculescu, A. M. Chiorcea-Paquim and S. H. P. Serrano, in *Electrochemical Sensors Analysis*, ed. S. Alegret and A. Merkoçi, Elsevier, Amsterdam, 2007, pp. 413–438.
60. F. Jelen, M. Tomschik and E. Palecek, *J. Electroanal. Chem.*, 1997, **423**, 141.
61. D. Marin, R. Valera, E. De la Red and C. Teijeiro, *Bioelectrochem. Bioenerg.*, 1997, **44**, 51.
62. D. Marin, P. Perez, C. Teijeiro and E. Palecek, *Biophys. Chem.*, 1998, **75**, 87.
63. V. Brabec, V. Vetterl and O. Vrana, in *Electroanalysis of Biomacromolecules. Experimental Techniques in Bioelectrochemistry, Bioelectrochemistry: Principles and Practice*, Vol. 3, ed. V. Brabec, D. Walz and G. Milazzo, Birkhauser Verlag, Basel, 1996, p. 287.
64. J. Wang, M. Chicharro, G. Rivas, X. H. Cai, N. Dontha, P. A. M. Farias and H. Shiraishi, *Anal. Chem.*, 1996, **68**, 2251.
65. V. Brabec, *Electrochim. Acta*, 2000, **45**, 2929.
66. O. Korbut, M. Buečková, P. Tarapčík, J. Labuda and P. Gründler, *J. Electroanal. Chem.*, 2001, **506**, 143.
67. J. Labuda, M. Buckova, S. Jantova, I. Stepanek, I. Surugiu, B. Danielsson and M. Mascini, *Fresenius J. Anal. Chem.*, 2000, **367**, 364.
68. R. Ovádeková, S. Jantová, S. Letašiová and J. Labuda, *Anal. Bioanal. Chem.*, 2006, **386**, 2055.
69. J. Labuda, K. Bubničová, L. Koval'ová and M. Vaníčková, *Sensors*, 2005, **5**, 411.
70. A. Ferancová, M. Adamovski, P. Gruendler, J. Zima, J. Barek, J. Mattusch, R. Wennrich and J. Labuda, *Bioelectrochemistry*, 2007, **71**, 33.
71. J. Wang, G. Rivas, M. Ozsoz, D. H. Grant, X. H. Cai and C. Parrado, *Anal. Chem.*, 1997, **69**, 1457.
72. G. Marrazza, I. Chianella and M. Mascini, *Anal. Chim. Acta*, 1999, **387**, 297.
73. G. Chiti, G. Marrazza and M. Mascini, *Anal. Chim. Acta*, 2001, **427**, 155.
74. M. Mascini, I. Palchetti and G. Marrazza, *Fresenius J. Anal. Chem.*, 2001, **369**, 15.
75. F. Lucarelli, I. Palchetti, G. Marrazza and M. Mascini, *Talanta*, 2002, **56**, 949.
76. A. M. Tencaliec, S. Laschi, V. Magearu and M. Mascini, *Talanta*, 2006, **69**, 365.

77. I. Szpakowska, B. Krassowska-Swiebocka, D. Maciejewska, P. Kazmierczak, W. Jemielita, M. Konrad, J. Trykowska and M. Maj-Zurawska, *Electroanalysis*, 2006, **18**, 1422.
78. M. Ozsoz, A. Erdem, P. Kara, K. Kerman and D. Ozkan, *Electroanalysis*, 2003, **15**, 613.
79. A. Mugweru and J. F. Rusling, *Electrochem. Commun.*, 2001, **3**, 406.
80. A. Mugweru and J. F. Rusling, *Anal. Chem.*, 2002, **74**, 4044.
81. N. Popovich and H. Thorp, *Interface*, 2002, **11**, 30.
82. Y. Zhang, H. Zhang and N. Hu, *Biosens. Bioelectron.*, 2008, **23**, 1077.
83. L. Havran, J. Vacek, K. Cahová and M. Fojta, *Anal. Bioanal. Chem.*, 2008, **391**, 1751.
84. A. M. Nowicka, A. Kowalczyk, Z. Stojek and M. Hepel, *Biophys. Chem.*, 2010, **146**, 42.
85. J. Labuda, M. Buckova, M. Vanickova, J. Mattusch and R. Wennrich, *Electroanalysis*, 1999, **11**, 101.
86. O. Korbut, M. Bučková, P. Tarapčík, J. Labuda and P. Gründler, *J. Electroanal. Chem.*, 2001, **506**, 143.
87. J. Labuda, R. Ovádeková and J. Galandová, *Microchim. Acta*, 2009, **164**, 371.
88. J. Wang, M. Ozsoz, X. H. Cai, G. Rivas, H. Shiraishi, D. H. Grant, M. Chicharro, J. Fernandes and E. Palecek, *Bioelectrochem. Bioenerg.*, 1998, **45**, 33.
89. J. Y. Gu, X. J. Lu and H. X. Ju, *Electroanalysis*, 2002, **14**, 949.
90. H. C. M. Yau, H. L. Chan and M. S. Yang, *Biosens. Bioelectron.*, 2003, **18**, 873.
91. X. Wang, T. Yang and K. Jiao, *Biosens. Bioelectron.*, 2009, **25**, 668.
92. X. Wang and K. Jiao, *Anal. Chim. Acta*, 2010, **664**, 34.
93. Y. Zhang and N. Hu, *Electrochem. Commun.*, 2007, **9**, 35.
94. M. Bučková, J. Labuda, J. Šandula, L. Križková, I. Štěpánek and Z. Ďuračková, *Talanta*, 2002, **56**, 939.
95. O. Korbut, M. Bučková, J. Labuda and P. Gründler, *Sensors*, 2003, **3**, 1.
96. J. Labuda, M. Bučková, L'. Heilerová, S. Šilhár and I. Štepánek, *Anal. Bioanal. Chem.*, 2003, **376**, 168.
97. J. Labuda, M. Bučková, L. Heilerová, A. Čaniová-Žiaková, E. Brandšteterová, J. Mattusch and R. Wennrich, *Sensors*, 2002, **2**, 1.
98. L. Heilerová, M. Bučková, P. Tarapcik, S. Šilhár and J. Labuda, *Czech. J. Food Chem.*, 2003, **21**, 78.
99. A. Ferancová, L. Heilerová, E. Korgová, S. Šilhár, I. Štěpánek and J. Labuda, *Eur. Food Res. Technol.*, 2004, **219**, 416.
100. D. Šimková, E. Beinrohr and J. Labuda, *Acta Chim. Slovaca*, 2009, **2**, 129.
101. L. D. Mello, S. Hernandez, G. Marrazza, M. Mascini and L. T. Kubota, *Biosens. Bioelectron.*, 2006, **21**, 1374.
102. B. Prieto-Simón, M. Cortina, M. Campàs and C. Calas-Blanchard, *Sens. Actuators, B*, 2008, **129**, 459.

103. A. M. Oliveira-Brett, J. A. P. Piedade and S. H. P. Serrano, *Electroanalysis*, 2000, **12**, 969.
104. P. A. Ropp and H. H. Thorp, *Chem. Biol.*, 1999, **6**, 599.
105. R. C. Holmberg, M. T. Tierney, P. A. Ropp, E. E. Berg, M. W. Grinstaff and H. H. Thorp, *Inorg. Chem.*, 2003, **42**, 6379.
106. F. C. Abreu, M. O. F. Goulart and A. M. O. Brett, *Biosens. Bioelectron.*, 2002, **17**, 913.
107. A. M. Oliveira Brett and L. A. da Silva, *Anal. Bioanal. Chem.*, 2002, **373**, 717.
108. J. A. P. Piedade, I. R. Fernandes and A. M. Oliveira-Brett, *Bioelectrochemistry*, 2002, **56**, 81.
109. A. M. Oliveira-Brett, A. M. Chiorcea Paquim, V. C. Diculescu and J. A. P. Piedade, *Med. Eng. Phys.*, 2006, **28**, 963.
110. A. M. Oliveira Brett and V. C. Diculescu, *Bioelectrochemistry*, 2004, **64**, 133.
111. A. M. Oliveira Brett and V. C. Diculescu, *Bioelectrochemistry*, 2004, **64**, 143.
112. M. A. La-Scalea, S. H. P. Serrano, E. I. Ferreira and A. M. O. Brett, *J. Pharm. Biomed. Anal.*, 2002, **29**, 561.
113. A. M. O. Brett, L. A. da Silva, H. Fujii, S. Mataka and T. Thiemann, *J. Electroanal. Chem.*, 2003, **549**, 91.
114. V. Vyskočil, J. Labuda and J. Barek, *Anal. Bioanal. Chem.*, 2010, **397**, 233.
115. S. C. B. Oliveira, O. Corduneanu and A. M. Oliveira-Brett, *Bioelectrochemistry*, 2008, **72**, 53.
116. L. Tian, W. Z. Wei and Y. Mao, *Clin. Biochem.*, 2004, **37**, 120.
117. J. F. Rusling, in *Electrochemistry of Nucleic Acids and Proteins. Towards Electrochemical Sensors for Genomics and Proteomics*, ed. E. Palecek, F. Scheller and J. Wang, Elsevier, Amstredam, 2005, pp. 433–449.
118. A. Mugweru, J. Yang and J. F. Rusling, *Electroanalysis*, 2004, **16**, 1132.
119. A. Mugweru, B. Wang and J. F. Rusling, *Anal. Chem.*, 2004, **76**, 5557.
120. M. So, E. G. Hvastkovs, J. B. Schenkman and J. F. Rusling, *Biosens. Bioelectron.*, 2007, **23**, 492.
121. G. Zauner, Y. Wang, M. Lavesa-Curto, A. MacDonald, A. G. Mayes, R. P. Bowater and J. N. Butt, *Analyst*, 2005, **130**, 345.
122. Z. Junhui, C. Hong and Y. Ruifu, *Biotechnol. Adv.*, 1997, **15**, 43.
123. C. K. O'Sullivan and G. G. Guilbault, *Biosens. Bioelectron.*, 1999, **14**, 663.
124. F. R. R. Teles and L. P. Fonseca, *Talanta*, 2008, **77**, 606.
125. J. Wang, M. Jiang and E. Palecek, *Bioelectrochem. Bioenerg.*, 1999, **48**, 477.
126. Y. Wu, L. Yi, Q. Xie, Y. Zhang, F. Yin and S. Yao, *Talanta*, 2001, **54**, 263.
127. Y. Liu, K. Deng, J. Li, S. Liu and S. Yao, *Biophys. Chem.*, 2004, **112**, 69.
128. L. Zhu, Y. Gao, H. Shen, Y. Yang and L. Yuan, *J. Anal. Chem.*, 2005, **60**, 877.

129. S. Takahashi, H. Matsuno, H. Furusawa and Y. Okahata, *Anal. Biochem.*, 2007, **361**, 210.
130. R. J. Rawle, M. S. Johal and C. R. D. Selassie, *Biomacromolecules*, 2008, **9**, 9.
131. A. M. Nowicka, A. Kowalczyk, Z. Stojek and M. Hepel, *Biophys. Chem.*, 2010, **146**, 423.

CHAPTER 8

# New Trends in DNA Sensors for Environmental Applications: Nanomaterials, Miniaturization, and Lab-on-a-Chip Technology

ALFREDO DE LA ESCOSURA-MUNIZ,[1] MARIANA MEDINA[1] AND ARBEN MERKOÇI[1,2]

[1] Nanobioelectronics & Biosensors Group, CIN2 (ICN-CSIC), Catalan Institute of Nanotechnology, Campus de la UAB, 08193 Bellaterra (Barcelona) Spain
[2] ICREA, Institució Catalana de Recerca i Estudis Avançats, 08010 Barcelona, Spain

## 8.1 Introduction

The permanent interest in monitoring pollutants in water, air, or elsewhere that may present risks to human health and the ecosystem suggests the development of diverse analytical techniques that may allow fast and cost-efficient detection methods.

The aim of this chapter is to summarize the latest trends in the use of nanomaterials (nanoparticles, quantum dots, nanotubes, *etc.*) as well as miniaturization and "lab-on-a-chip" technologies for nucleic acid (NA)-based biosensing systems with interest for environmental applications.

The emergence of nanoparticles (NPs) as an alternative to other labels (fluorescent dyes, enzymes, *etc.*) for DNA detection has been previously reviewed by our group.[1] In addition to NPs, other materials such as carbon nanotubes (CNT) have been also reviewed for their general applications in sensors[2] and analytical methods[3] in general, with some examples of DNA

---

Nucleic Acid Biosensors for Environmental Pollution Monitoring
Edited by Marco Mascini and Ilaria Palchetti
© Royal Society of Chemistry 2011
Published by the Royal Society of Chemistry, www.rsc.org

detection systems. Although most of the principles of the nanomaterial-based sensors have been addressed previously, in this chapter we aim to focus on DNA and nanomaterial-based systems that are of interest for environmental applications.

In addition to nanomaterials, the chapter will also consider lab-on-a-chip systems involving DNA as a novel trend for environmental monitoring. Electrochemically based lab-on-a-chip devices were reviewed earlier by our group,[4] but here we focus on some applications related to the current use and future applications of these systems for environmental monitoring. Because of the lack of reports on specific DNA-based systems for some of the detection modes under review, other analytes will be also discussed keeping in mind future extension of these methods as DNA-based environmental monitoring possibilities.

## 8.2 Nanomaterial-based Sensors for DNA Detection

The discovery and study of nanomaterials has enabled the development of ultrasensitive optical and electrochemical biosensors, because of their large surface area, favorable optical and electronic properties, and electrocatalytic activity as well as good biocompatibility due to their nanometer size and specific physicochemical characteristics. The resulting sensors have been applied in areas of food quality, clinical analysis, and environmental control.

Different detection approaches for pathogen-related DNA are reviewed in this section. In most cases, the analysis of environmental pollution using nanomaterial-based biosensors consists of the detection of DNA related to environmental pathogens, using single-stranded (ss) DNA as a bioreceptor and final optical or electrochemical transducing. Nanomaterials can be used to modify the surface of the transducer, improving the performance of the sensor, or as optical or electroactive labels for the sensitive detection of pollutants.

### 8.2.1 Optical Sensors

#### 8.2.1.1 Nanoparticles as Optical Labels

The intrinsic optical properties (UV–visible light absorption and autofluorescence properties) of nanoparticles (NPs) and also their ability to change optical properties of sensor surfaces (*i.e.*, in surface plasmon resonance and devices based on light scattering) have been applied for their detection when used as labels in DNA hybridization assays.

It is well known, for example, that the NP plasmon band shifts when gold NPs (AuNPs) are aggregated, owing to a decrease in the interparticle distance. The absorbance maximum shifts from 520 nm to 580 nm (red to blue) when the aggregation takes place, and this phenomenon can be monitored for biosensing purposes. For example, an AuNP suspension previously modified with a ssDNA probe has been hybridized with a target DNA (complementary at its

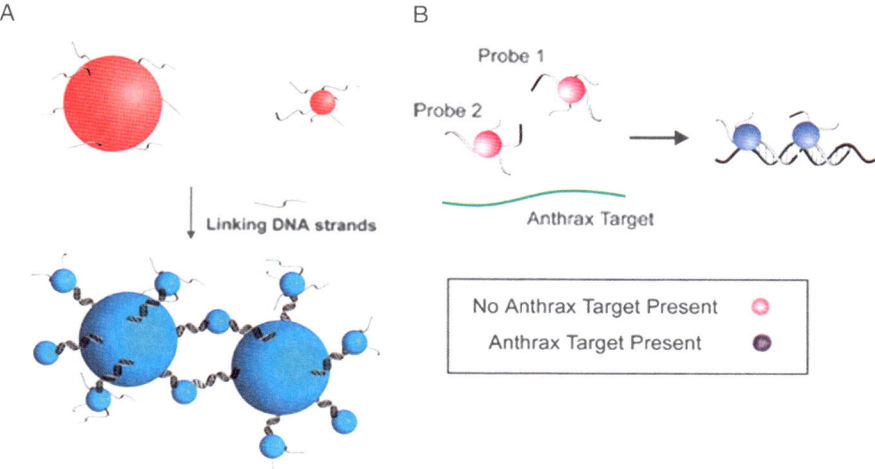

**Figure 8.1** (A) General scheme of the principle of the DNA detection based on the change in the absorption spectra due to the AuNPs aggregation. (B) Application for the detection of an anthrax PCR product. Adapted from ref. 5, with permission.

two ends with ssDNA), thus giving rise to the aggregation of NPs with the consequent change in color (Figure 8.1). This principle was pioneered by Mirkin's group and applied to the detection of DNA characteristic of anthrax, a biological warfare agent.[5]

Silica NPs (SiNPs), doped with either magnetic materials or dyes, have also been widely exploited for the collection, purification and detection of DNA/RNA.[6] The bioconjugation of SiNPs with DNA/RNA molecules can provide unique biofunctions. These NPs tend to exhibit very high luminescence and photostability, within a broad size range from 5 to 400 nm. Furthermore, dye-doped SiNPs have been extensively used in bioimaging and biochemical analysis because of advantages such as signal enhancement, photostability, and surface modification availability for the immobilization of biomolecules.[7] For example, SiNPs were doped with an europium ternary complex by the reverse microemulsion method[8] and used as fluorescence labels for the detection of *Escherichia coli* in water samples, resulting in an effective tool for environmental monitoring of pathogen DNA.

Core-shell NPs such as magnetic/luminescent $Fe_3O_4$/Eu:$Gd_2O_3$ synthesized by spray pyrolysis were used by Son *et al*.[9] both as platforms for the hybridization reactions and as fluorescence labels. The DNA hybridization sensor they developed has the capability of distinguish perfectly matching targets from two-base pair mismatching, allowing the discrimination of different bacterial species. These results opened the way to the quantification of specific bacteria based on the 16 S ribosomal DNA (rDNA) gene sequence in environmental samples. Moreover, core-shell NPs are highly versatile because they can be synthesized in many colors with a variety of phosphorescent lanthanide ions (*e.g.*, europium, terbium, samarium), allowing a multiplex detection of a number of targets via optical encoding.

The autofluorescence properties of inorganic semiconductor nanocrystal CdSe/ZnS core-shell NPs were applied by Vora *et al.* in DNA microarrays for the sensitive detection of pathogenic organisms from environmental matrices, allowing the detection of $10^3$ polymerase chain reaction (PCR) copies of *Vibrio cholera*.[10] In the same work, the resonance light scattering properties of AuNPs were also used for the detection of a few copies of the PCR product.

Other optical properties of NPs that have been utilized for DNA hybridization detection and are promising for their further application in environmental monitoring are the ability of AuNPs to act as quenchers of the fluorescence of other markers,[11] and the ability of AuNPs to change the scattering properties of a surface where they are attached through the hybridization reaction.[12]

### 8.2.1.2 Nanomaterials as Modifiers of Optical Transducers

In addition to their use as optical labels, NPs can be used as modifiers of the optical transducer surface, improving the performance of the optical sensor. A NP-modified surface for the optical detection of DNA has been reported by Endo *et al.*[13] It consisted of a gold-capped NP layer substrate immobilized with peptide nucleic acids (PNAs) for the development of a localized surface plasmon resonance (LSPR)-based label-free optical biosensor (Figure 8.2). The NP layer was formed on the gold-deposited glass substrate by the surface-modified SiNPs using a silane-coupling reagent. The PNAs recognize the DNA sequence of interest and the PNA–DNA hybridization produces a change in the LSPR signal that allows detection both of target oligonucleotides and of PCR-amplified real samples. This detection mode may be applied in the future for the environmental control of pathogens.

In addition to NPs, one-dimensional (1D) nanostructures (including nanowires, nanotubes, nanobelts, and nanorods) have also received significant research attention over the past years, due to their intriguing properties and promising applications.[14] In particular, porous structures of these 1D nanomaterials, owing to their intrinsic pore structures and high surface-to-volume ratio, have been the focus of recent studies, which further broaden the potential applications in catalysis, bioengineering, environmental protection, sensors, and other areas. Their advantageous properties have also been applied for use as modifiers of optical transducers. For example, layered-lanthanum crystalline nanowires (NWs) with hierarchical pores were synthesized via a hydrothermal route by Wang *et al.*[15] and luminescence properties were achieved by doping nanowires with $Eu^{3+}$. By combining the merits of hierarchical porous nanowires and layered hydroxides, these products have shown a unique bioengineering application of capture and rapid release of short DNA fragments in dilute solution. Environmental engineering applications resulting

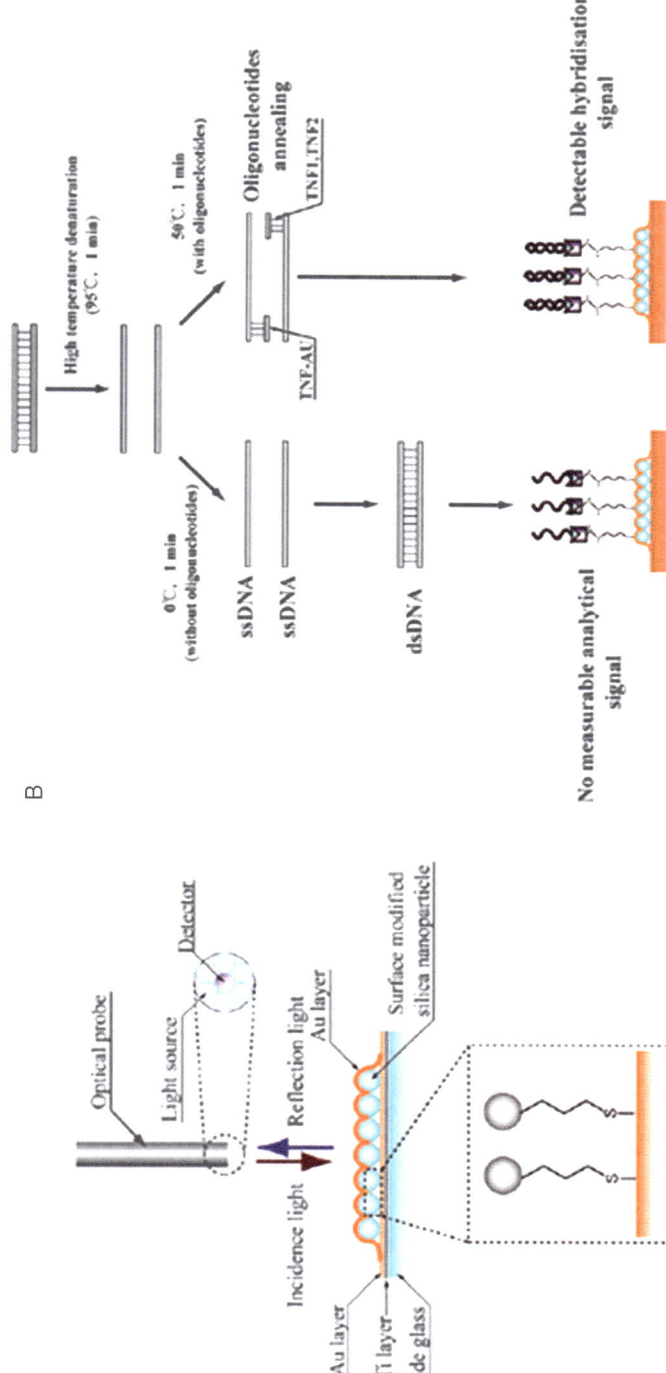

**Figure 8.2** (A) Experimental setup and construction of LSPR-based optical biosensor. (B) Label-free detection of PCR-amplified samples using LSPR-based optical biosensor. Adapted from ref. 13. with permission.

from the remarkable capability of these products to remove an organic dye (Congo red, a common azo dye in the textile industry) from wastewater have been previewed.

### 8.2.2 Electrochemical Sensors

Electrochemical biosensors based on immobilized DNA combine detection sensitivity with a high specificity for biomolecules. This reduces the consumption of DNA and gives rise to the development of modern methods of analysis of chemicals affecting DNA, including toxic substances with environmental and biological effects.[16,17] Various electrochemical DNA systems for environmental analysis have been developed.

#### 8.2.2.1 Nanoparticles as Electroactive Labels

The excellent electroactivity of metallic NPs, together with their easy bioconjugation, has given rise to their extensive use as labels in DNA sensors in recent years. Several electrochemical routes (voltammetric, potentiometric, conductometric, impedimetric, and scanning electrochemical microscopy methods) have been exploited for the sensitive detection of these NPs tags in bioassays.[18] Although in most cases these electrochemical DNA biosensors have not yet been used for environmental monitoring, the established methodologies may open the way for very sensitive, simple, and cost-efficient sensors for future applications.

NP-based amplification schemes have improved the sensitivity of bioelectronic assays by several orders of magnitude. In 2001, the groups of Wang[19] and Limoges[20] both pioneered the use of AuNP tags for electronic detection of DNA hybridization. This protocol relies on capturing the NPs to the hybridized target, followed by highly sensitive anode-stripping electrochemical measurement of the metal tracer.

Commonly, coupling the biorecognition element to the surface of magnetic beads effectively minimizes nonspecific binding. The hybridization of probe-coated magnetic beads with metal-tagged targets results in three-dimensional network structures of magnetic beads, cross-linked together through DNA and AuNPs. Such a magnetic-bead/DNA/metal-label assembly packing onto the electrode leads to direct contact between the metal label and the surface, and enables electrochemical measurements without dissolving the metal tag.[21-23] Electronic DNA hybridization assays have been extended to other metal NP tracers, including AgNPs,[24] CdS QDs,[25] and $Fe_2O_3$/Au core-shell NPs,[26] opening the way to analysis for environmental control.

Furthermore, the well-known catalytic properties of AuNPs on the reduction of silver ions have been applied by Lee *et al.*[27] for the sensitive detection of ssDNA in sandwich assays on ITO electrodes, offering a protocol that has great potential for simple, reproducible, highly selective, and sensitive DNA detection on fully integrated microdevices in environmental monitoring applications.

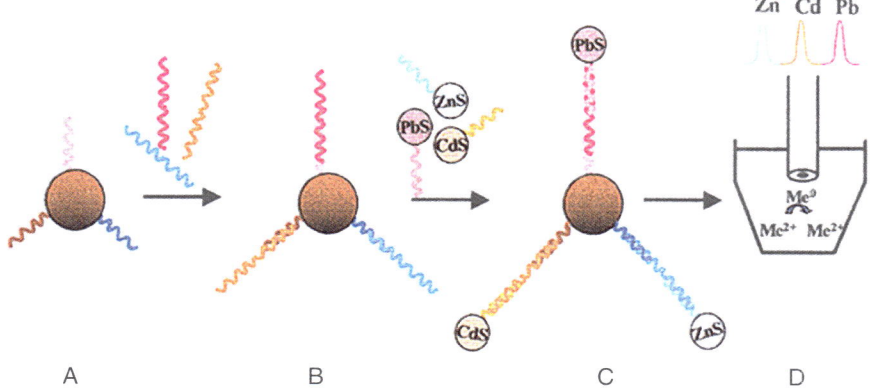

**Figure 8.3** Multitarget electrical DNA detection protocol based on different inorganic colloid nanocrystal tracers. (A) Introduction of probe-modified magnetic beads. (B) Hybridization with the DNA targets. (C) Second hybridization with the QD-labeled probes. (D) Dissolution of QDs and electrochemical detection. From ref. 28, with permission.

Other strategies based on the use of inorganic nanocrystals offering an electrodiverse population of electrical tags as needed for designing electronic coding have been reported (Figure 8.3). Three encoding NPs (ZnS, CdS, and PbS) have been used in this way to differentiate the signals of three ssDNA target strands in connection with a sandwich hybridization assay and stripping voltammetry of the corresponding metals.[28]

Finally, we can also highlight a variation in the use of NPs as direct electroactive labels, which consists in their use as carriers of a large quantity of other labels with electrochemical properties, achieving an improvement in the sensitivity of the assays. For example, a simple, sensitive electrochemical DNA biosensor based on *in situ* DNA amplification with AgNPs as carrier of horseradish peroxidase (HRP) electroactive labels has been designed by Fu *et al.*[29] Other nanomaterials such as CNTs have also been used as carriers. For example, the use of CNTs as carriers for several thousand enzyme tags has been reported, allowing the detection of DNA down to 1.3 zmol,[30–32] with future potential for environmental monitoring.

### 8.2.2.2 Nanomaterials as Modifiers of Electrotransducers

The presence of NPs on the electrotransducer surface promotes electron transfer, improving the electrochemical responses while using potentiometric and conductimetric techniques. Furthermore, some NPs provide a congenial microenvironment, similar to that of redox proteins in a native system, for retaining their bioactivity, giving the molecules more freedom in orientation. For these reasons, attempts to develop DNA hybridization assays using nanostructurated surfaces have been reported in the last few years.

The introduction of NPs into the transducing platform is achieved by their adsorption onto conventional electrode surfaces in various forms, including composites.

Regarding NP application for DNA detection of environmental interest, Feng et al.[33] constructed AuNP/polyaniline nanotube membranes on a glassy carbon electrode (Au/nanoPAN/GCE) for sensing DNA. The properties of the Au/nanoPAN/GCE and the characteristics of the immobilization and hybridization of DNA were studied by cyclic voltammetry (CV), differential pulse voltammetry (DPV), and electrochemical impedance spectroscopy (EIS). The synergistic effect of the two nanomaterials, AuNP and nanoPAN, dramatically enhanced the sensitivity for DNA hybridization recognition, achieving a detection limit of $3.1 \times 10^{-13}$ mol L$^{-1}$ of a specific DNA sequence of the phosphinothricin acetyltransferase gene.

Other nanomaterials used as modifiers of electrotransducers are 1D NWs. They have been used for bridging two closely spaced electrodes for label-free DNA detection. A p-type silicon nanowire, functionalized with PNA probes, was shown to be extremely useful for real-time, label-free, conductimetric monitoring of the hybridization event.[34] This technique relies on the binding of the negatively charged DNA target, which leads to an increase in conductance, reflecting the increased surface charge.

Hybrid techniques (mechanic/electric), such as microgravimetric methods based on piezoelectric quartz crystal (PQC) sensing, are also very sensitive methods for the detection of DNA hybridization. Compared to electrochemical and optical methods, the PQC biosensor provides a label-free detection method using a simple and portable equipment set-up, hence having the potential for direct detection of environmental pathogens. A practical approach to enhance the sensitivity of PQC biosensors consists of incorporating NPs onto biosensing layers in order to increase the amount of active recognition molecules at the surface of the PQC biosensor and facilitate the capture of the DNA analyte. In this way, a highly sensitive PQC DNA biosensor was fabricated by photodepositing silver NPs (AgNPs) modified with neutravidin molecules on $TiO_2$-coated PQC electrodes.[35] The immobilizaton of biotinylated probe DNA strands is greatly improved by the NPs, making it possible to achieve a limit of detection of 0.4 ng of PCR products related to *E. coli* in 1 L of drinking-water.

## 8.3 DNA and Nanomaterial-based Sensing Platforms

The structure of DNA is very sensitive to the influence of environmental pollutants (*e.g.*, heavy metals, polychlorinated biphenyls, or polyaromatic compounds). These substances are characterized by a great affinity to DNA, causing mutagenesis and carcinogenesis. Based on these characteristics it is very attractive to use DNA-containing systems (*e.g.*, DNA-based biosensors) to perform genotoxic assays, or for rapid testing of pollutants for mutagenic and carcinogenic activity. Furthermore, some heavy metals have the ability to coordinate with DNA bases, allowing their capture by DNA probes and later detection.

## 8.3.1 Optical Detection Methods

The coordination-based interaction between $Hg^{2+}$ and bisthymine has recently attracted significant interest.[36] In detail, T–T mismatches in DNA duplexes selectively and strongly capture $Hg^{2+}$ (binding constant higher than A–T), and the metal-mediated T–Hg–T forms stable DNA duplexes. This principle can be applied to selectively bind $Hg^{2+}$, which can be detected later following different strategies where NPs can be used as labels or as enhancers of the analytical signal.

An example of this approach has recently been reported by Wang et al.[37] They designed a visual and fluorescent sensor for $Hg^{2+}$ in aqueous solution, based on the $Hg^{2+}$-induced conformational change of a T-rich ssDNA and the difference in electrostatic affinity between ssDNA and dsDNA with AuNPs (Figure 8.4). The dye-tagged ssDNA containing T–T mismatched sequences was chosen as $Hg^{2+}$ acceptor. At high ionic strength, introduction of the ssDNA to a colloidal solution of the aggregates of AuNPs results in a color change of the solution, from blue-gray to red, and the fluorescence quenching of the dye. Binding of $Hg^{2+}$ by the ssDNA induces the double-stranded structure formation. This dsDNA formation reduces the capability to stabilize bare NPs against salt-induced aggregation, the color of the solution remaining blue-gray, but the fluorescence signal enhancement compared to that without $Hg^{2+}$ achieved a detection limit of 40 nM. Furthermore, both the color and fluorescence changes of the system were extremely specific for $Hg^{2+}$ even in the

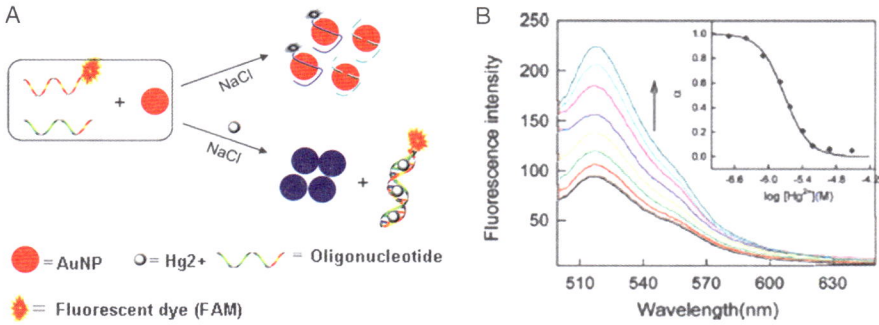

**Figure 8.4** (A) Schematic description of colorimetric and fluorescent sensing of $Hg^{2+}$ based on the modification-free AuNPs. (B) Changes in the fluorescence emission spectra of the dye-DNA modified AuNPs in the Tris-HCl buffer (0.1 M NaCl) upon addition of different concentrations of $Hg^{2+}$. Excitation wavelength was 480 nm. Inset: Response (R) parameter as a function of logarithm of $Hg^{2+}$ concentration (M) at pH 7.4. From ref. 37, with permission.

presence of high concentrations of other heavy metal and transition metal ions, which meet the selective requirements for environmental applications.

The same principle has recently been applied by Liu et al. for $Hg^{2+}$ detection at 1.0 nM levels in water samples[38] by measuring the enhanced light-scattering plasmon resonance signals resulted from $Hg^{2+}$–DNA complex-induced aggregation of AuNPs.

*DNAzymes* (DNA-based biocatalysts capable of performing chemical transformations) have also been used for heavy metal detection, taking advantage of NP properties. In this way, Wei et al.[39] reported a simple, sensitive, and label-free DNAzyme-based sensor for $Pb^{2+}$ detection using unmodified AuNPs. The technique is based on the fact that unfolded ssDNA can be adsorbed on the citrate-protected AuNPs while dsDNA cannot. Using this method the substrate cleavage by the DNAzyme in the presence of $Pb^{2+}$ can be monitored by color change of AuNPs. $Pb^{2+}$ detection was realized with a detection limit of 500 nM. A similar approach was reported by Lu's group[40] for the detection of 100 nM $Pb^{2+}$.

In addition to heavy metals, other pollutants have also been optically detected using this kind of DNAzyme biosensor. For example, colorimetric uranium sensors based on uranyl ($UO_2^{2+}$)-specific DNAzyme and AuNPs have been developed and demonstrated using both labelled and label-free methods for the sensitive detection of this ion, the most soluble and bioavailable form of uranium (Figure 8.5).[41]

In the labelled method, a uranyl-specific DNAzyme is attached to the AuNPs, forming purple aggregates. The presence of the uranyl ion induces disassembly of the DNAzyme-functionalized AuNP aggregates, resulting in red individual AuNPs and giving rise to a "turn-on" sensor that can detect 50 nM of uranyl. On the other hand, the label-free method utilizes the different adsorption properties of ssDNA and dsDNA on AuNPs, which affects the stability of AuNPs in the presence of NaCl. The presence of uranyl results in cleavage of substrate by DNAzyme, releasing ssDNA that can be adsorbed on AuNPs and protects them from aggregation. Taking advantage of this phenomenon, a "turn-off" sensor was developed with a detection limit of 1 nM. Moreover, both sensors showed excellent selectivity over other metal ions.

*Aptasensors* represent another kind of DNA-based biosensors that has been applied for environmental monitoring. Aptamers are single-stranded DNA or RNA ligands that can be selected for different targets, starting from a huge library of molecules containing randomly created sequences. DNAzymes can only perform chemical modifications on NAs, whereas aptamers can bind a broad range of molecules. A combination of the two has generated a new class of functional NAs known as allosteric DNAzymes or *aptazymes*,[42] the characteristics of which can be improved using nanomaterials. For example, an adenosine-dependent aptazyme built on the basis of the $Pb^{2+}$-specific DNAzyme previously described was used to assemble AuNPs.[43] In the presence of adenosine, the substrate is cleaved and the assembly inhibited. Several aptazymes (activated RNAzymes) have been shown to have metal-ion-dependent activities,[44] with further environmental monitoring interest.

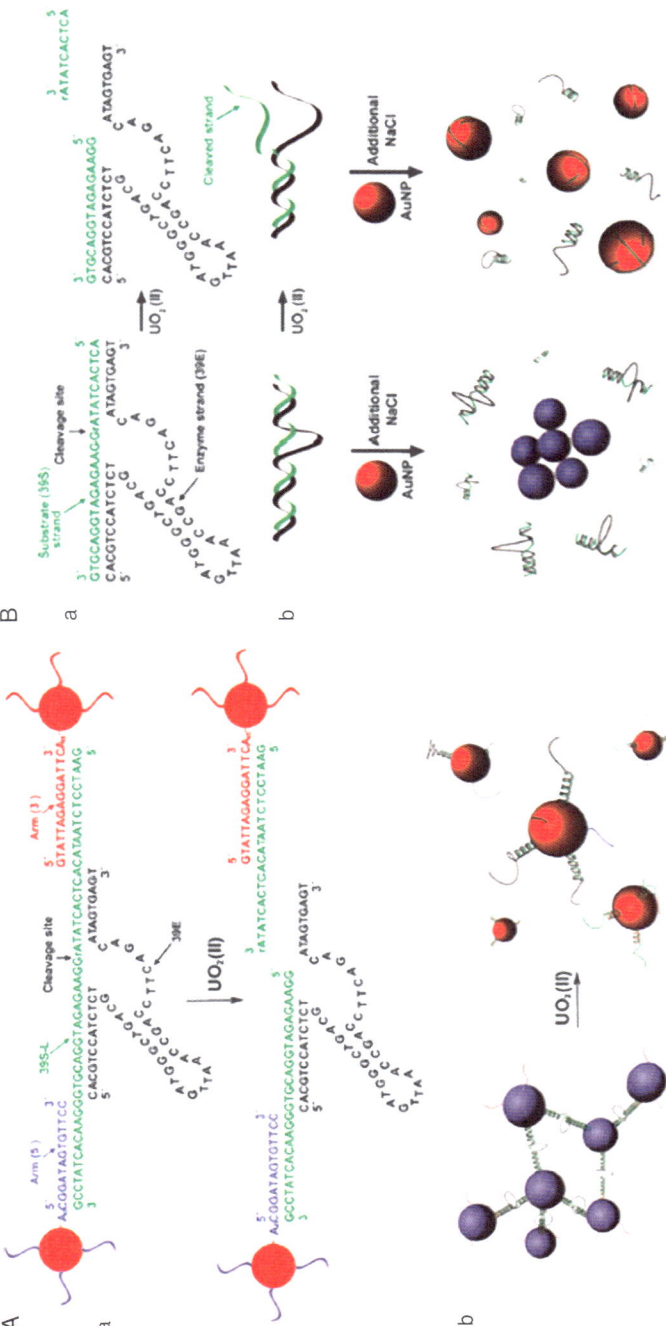

**Figure 8.5** (A) (a) Scheme of labelled colorimetric sensor based on AuNP disassembly in the absence and presence of $UO_2^{2+}$. In the presence of $UO_2^{2+}$, the length of the weakest complementary part in the aggregates becomes shorter due to $UO_2^{2+}$-induced substrate cleavage. The substrate cleavage can decrease the melting temperature of AuNP aggregates. (b) As $UO_2^{2+}$ is introduced into AuNP aggregates and the temperature is controlled above the melting temperature of $UO_2^{2+}$-treated aggregates, AuNP disassembles. (B) (a) Design and sequence of the label-free sensor (complex). After $UO_2^{2+}$-induced cleavage, 10 mer ssDNA is released which adsorbs on AuNP surface. (b) AuNP reaction in addition of $UO_2^{2+}$ treated/untreated complex and additional NaCl. AuNPs aggregate in the absence of $UO_2^{2+}$ but remain dispersed in its presence. From ref. 41, with permission.

## 8.3.2 Electrochemical Detection Methods

The previously explained strong interaction between $Hg^{2+}$ and bisthymine has also been investigated using electrochemical sensors. Zhu et al.[45] recently reported a sensor for the highly sensitive detection of $Hg^{2+}$ in aqueous solutions, using a T-rich, mercury-specific oligonucleotide (MSO) probe and AuNP-based signal amplification (Figure 8.6). The MSO probe contains seven T bases at both ends and a "mute" spacer in the middle, which, in the presence of $Hg^{2+}$, forms a hairpin structure via the $Hg^{2+}$-mediated coordination of T–$Hg^{2+}$–T base pairs. A thiolated MSO probe is immobilized on gold

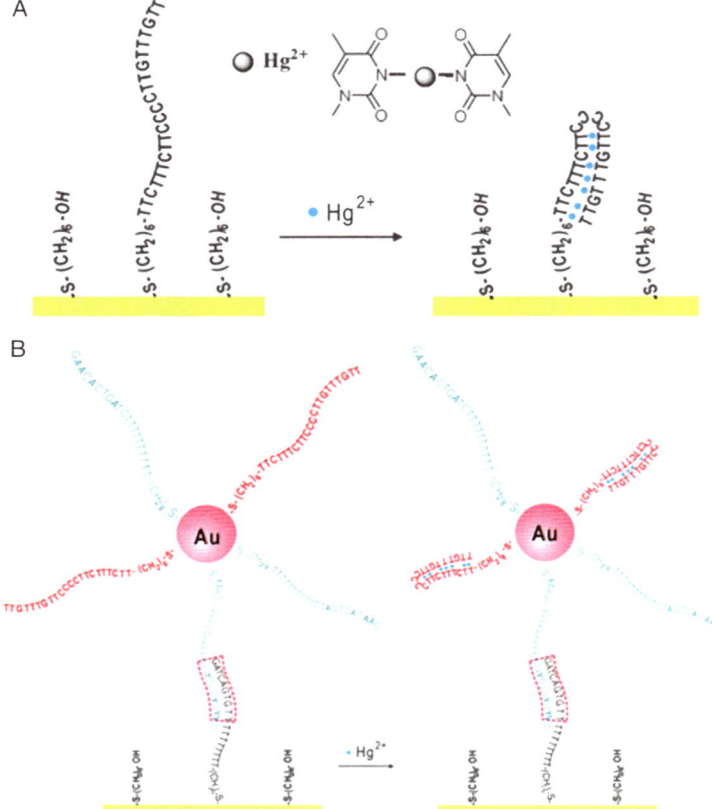

**Figure 8.6** Scheme of the two modes of a sensor for the highly sensitive detection of $Hg^{2+}$ in aqueous solutions, using a thymine (T)-rich, mercury-specific oligonucleotide (MSO) probe and AuNPs-based signal amplification. (A) Directly immobilized MSO probe. Inset corresponds to the hairpin structure via the $Hg^{2+}$-mediated coordination of T–$Hg^{2+}$–T base pairs formed by MSO probe in the presence of $Hg^{2+}$. (B) AuNPs mediated immobilized MSO probe. From ref. 45, with permission.

electrodes to capture free $Hg^{2+}$ in aqueous media, and the MSO-bound $Hg^{2+}$ can be electrochemically reduced to $Hg^+$, which provides a readout signal for quantitative detection of $Hg^{2+}$. This direct immobilization strategy leads to a detection limit of 1 μM. This sensitivity is improved by using MSO probe-modified AuNPs to amplify the electrochemical signals. AuNPs are comodified with the MSO probe and a linking probe that is complementary to a capture DNA probe immobilized on gold electrodes, giving an amplification factor of more than three orders of magnitude. This leads to a limit of detection of 0.5 nM (100 ppt) with an excellent selectivity over a spectrum of interference metal ions.

Zheng et al.[46] have reported a CNT-based DNA biosensor for the sensitive monitoring of phenolic pollutants. The biosensor was constructed by immobilizing DNA on a GCE modified with multiwall carbon nanotubes (MWCNTs) dispersed in Nafion (DNA/MWCNTs/GCE). The DNA-modified electrode exhibited two well-defined oxidation peaks corresponding to the guanine (G) and adenine (A) residues of DNA, respectively. Phenol, m-cresol, and catechol showed noticeable inhibition towards the response of the electrode due to their interactions with DNA. These findings were used to design biosensors with linear response to these phenolic pollutants, obtaining detection limits of 16 μM, 1.2 μM, and 0.52 μM for phenol, m-cresol, and catechol respectively.

In spite of their use as sensing material, the presence of CNTs in the environment can itself represent an environmental contamination and consequently there is also a demand for CNT detection sensing systems. For this reason Peng et al.[47] developed an electrochemical DNA sensor that can distinguish single-walled CNTs (SWCNTs) and MWCNTs both in buffer and in cell extracts. It is based on the capability of SWCNTs to specifically induce the formation of human telomeric i-motif (a tetrameric DNA structure with protonated cytosine–cytosine base pairs). This sensor can selectively detect SWCNTs with a detection limit of 0.2 ppm and its utility has been demonstrated in cancer cell extracts. The DNA sensor consists of a short ssDNA containing a human telomeric i-motif sequence. This DNA is modified with a redox-active dye, methylene blue (MB), and attached to gold electrodes. In the absence of target, the immobilized 26-mer DNA remains unfolded in the buffer, observing a faradaic current due to the attached MB tag. In the presence of the i-motif cDNA, the faradaic current decreases, indicating the formation of telomeric DNA duplex, which prevents the MB tag approaching the electrode surface, allowing the detection of 0.2 ppm of SWCNTs (Figure 8.7). Furthermore, the decrease in the signal strongly depends on the type of CNT, making possible the discrimination between SWCNTs and MWCNTs. This methodology seems to provide new insights into how to design a biosensor for CNT detection.

## 8.4 Nanoprobe- and Nanochannel-based Sensing

The use of CNTs as a probe in scanning probe microscopy has become an interesting alternative for real applications with industrial interest. The unique mechanical buckling properties of the CNT seem to diminish the imaging force

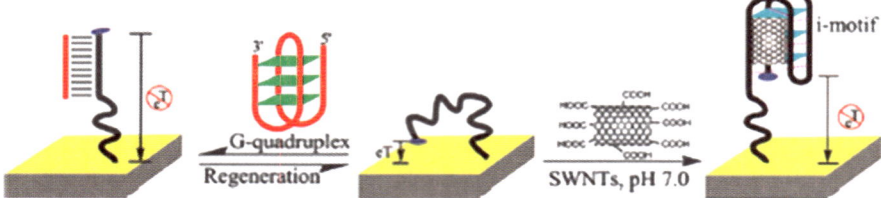

**Figure 8.7**  Schematic of the i-motif telomeric DNA-based electrochemical sensor for the detection of SWCNTs. The sensor is based on a conformation change of an electrode-bound, methylene blue-modified human telomeric C-rich sequence. The plain and dashed lines show the hydrogen bonding formed by the C·C+ hemiprotonated base pair of the "building blocks" for i-motif formation on right panel. From ref. 47, with permission.

exerted on the sample, thus making CNT scanning probes ideal for imaging soft materials, including samples in liquid environments. Stevens et al.[48] demonstrated the instability of CNT scanning probes when submerged in aqueous solution and introduced a novel approach to resolve this chemical incompatibility by coating the CNT probe with ethylenediamine, thus rendering the CNT probe less hydrophobic and showing the liquid imaging capability of treated CNT probes. This set-up allows the imaging of DNA molecules in aqueous solutions, opening the way to further environmental applications.

The use of synthetic nanopores/nanochannels for biosensing also shows promise. In natural ion channels the ionic current flows through and is altered when a molecule binds to a specific region of the channel. This principle is currently being applied for biosensing purposes, by using biological ($\alpha$-hemolysin protein) or synthetic (alumina or silicon nitride membranes) biomimetic nanopores and nanochannels, simulating this natural behavior. Nanoporous materials also show a dramatic increase in surface/volume ratio that enhances the signals corresponding to interaction between solutes and surfaces, including biomolecular reactions. In the case of DNA hybridization detection, the sensing principle is based in the fact that if a ssDNA is immobilized inside the channel, hybridization with the specific target produces a change in the ionic current through the channel that can be measured and related to the DNA concentration.[49] Based on these principles, nanopore/nanochannel arrays and single nanopores seem to present promising new features for the development of DNA sensors with potential for environmental monitoring.

## 8.5  Lab-on-a-chip Systems

The current trend in the development of biosensing systems is towards the miniaturization of the whole analytical chain, from sampling to the detection of analytes. The advantages of miniaturized systems in general, and of microfluidic-based biosensors in particular, include reduction of the amount of sample and reagents, detection facility, minimal handling of hazardous materials, and multiple and parallel sample detection capability. All of these characteristics make the process of environmental pollution and toxicity monitoring more efficient and cheaper.

Micro total analysis systems (μTAS) and lab-on-a-chip (LOC) platforms have been available for several years and shown to be an interesting field of research and applications. Their size characteristics and production facilities allow the handling of extremely small quantities of fluids (picoliters), representing a promising option to improve the detection limit and sensitivity of NA-based biosensors.

Some approaches have been proposed for on-chip amplification and analysis of NAs. These devices can be used to detect DNA or RNA in various kinds of samples, thanks to their selectivity and wide range of physical, chemical, and biological activity.[50]

Some lab-on-a-chip designs, their constituent materials and fabrication processes, control systems, and interfaces with nanoscale/nanomaterial based biosensors for future environmental monitoring possibilities are summarized in the following sections.

### 8.5.1 Fabrication Technologies

#### 8.5.1.1 Micromanufacturing Techniques

Microtechnology originated in the 1960s, when academics and researchers were interested in developing chips containing many transistors in a small area, which remarkably improved performance and functionality, as well as decreasing costs and the volumes necessary for analysis, giving rise to what was known as the computer science revolution.[51] The term *microelectromechanical system* (MEMS) was coined in the 1980s to describe the new mechanical systems on a chip, such as microelectrical resonators and motors, among others. MEMS processing is generally divided into surface mechanized processes and processes that generate more complex 3D geometries. MEMS technology has generated many products in areas like the automotive industry, printing, electronics and computer science, among others.[52] MEMS applications are of special interest for biosensing applications. Several micromanufacturing techniques, as described below, have been developed and applied for chip fabrication.

- *Photolithography*. This technique consists of transferring a pattern from a photomask to the surface of a silicon wafer in crystalline form. This wafer is then used as a substrate. It is possible to use other substrates, such as glass, polymers, sapphire and metals, among others.[53] 3D structures fabricated using this process have potential applications in MEMS sensors/actuators, optical devices, and microfluidics.[54]
- *Soft lithography* is related to the manufacturing techniques of microstructures, using elastomers, molds, and conformable photomasks. It is called "soft" because elastomeric materials are used, especially poly-(dimethilsiloxane) (PDMS). This technique is generally used to build

devices at the micro- and nanometer scale. Processes include impression by micro contact, molding by replica, molding by micro transference, and micromolding in capillaries.[55]

- *Micromachining* is an important manufacturing technology. It is based on the use of a cutting tool to obtain predefined material geometries.[56] Although the sizes obtained with micromachining are not as small as those obtained using lithography, this technique allows the fabrication of platforms for integration with smaller devices. Other processes used to fabricate microfluidic systems are X-ray lithography,[57] high-precision molding, hot embossing,[58] microinjection molding,[59] and roll-to-roll embossing.[60] This technology has great potential, thanks to the facility of designing complex 3D structures in a great variety of materials, such as metals, polymers, glass, and ceramics.

- Finally, *microerosion processes* use poly(dimethylsiloxane) (PDMS) and flexopolymer as masking materials for the 3D microstructuring of glass, using powder blasting technology. Such elastomeric masks provide a promising alternative to metallic contact masks for the definition of microstructures[61] and laser micromachining.[62]

*Ink-printing technology* was initially used for computer science and decorative objects. Now it is widely used to print on different surfaces, such as aluminum, glass, plastic, and paper, using special inks for different purposes like graphite inks, silver, CNT inks, and polymer inks, among others. It allows printing geometries with certain degrees of thickness and roughness to create 3D structures, without the use of masks.[63]

### 8.5.1.2 Sample Treatment and Transport

The study of microfluids is an interdisciplinary field, where sciences like physics, chemistry, engineering, and biotechnology converge. Microfluidics analyzes the behavior of fluids at the microsopic scale, which differs considerably from their macrocospic behavior; specifically, characteristic such as surface tension and energy dissipation change considerably. In microchannels (diameter 10–500 nm), the Reynolds number is extremely low; this is why the flow is always laminar and diffusion phenomena take place in fluid mixtures.[64]

Microfluidics is related to systems and methods that manipulate fluids on the micron scale. Aspects that must be considered include (1) dimensions of the components; (2) geometry (length, cross-sectional section of the channels, and properties of the surface); (3) characteristics of the surface; (4) characteristics of the fluid (density, viscosity, electrical and thermal conductivity, diffusion coefficients, surface tension, among others), and (5) ambient temperature.

Microfluidic design depends on the application. In cases of special operational requirements such as bioanalytical reactions or biological sample treatment, the design of microfluidic geometries plays an important role. Sample transport and processing can be performed in different ways, such as: (1) electrophoresis;[65,66] (2) dielectrophoresis;[67] (3) acoustic and magnetic stimuli;[68] (4) pneumatic injection;[69] (5) pressure difference between inlet and outlet, among others. Various detection systems (optical, magnetic, acoustic, electrochemical) have been reported.

### 8.5.1.3 Integration with Nanoscale Biosensing

DNA-based biosensors and molecular technologies such as the PCR for detection of pathogens are slowly replacing culture-based detection methods, because their higher sensitivity and selectivity and lower times of analysis than culture-based methods. Scientists are overcoming this limitation by miniaturizing molecular assays into a microsystems format. Microfluidics is the most useful technology for the development of miniaturized devices that transport, mix, and control reactive fluid volumes in the micron range.[70]

The use of electronics principles to make the interface between bionanosensors with the microsize platform is continuously increasing because of its potential for easy integration. Microfluidic-based electrochemical sensors, using CNTs and paramagnetic microparticles,[71] NWs,[72] CNT transistors,[73] and field-effect transistor (FET) sensors[74] have been reported in recent years.

Finally, problems such as nonspecific absorption of the sample in the biosensors and total integration of systems are being widely studied.[75] New challenges in detection systems, such as the use of new materials or some other modified materials to allow the integration of micro- and nanoscale devices, are being proposed.

## 8.5.2 Operation

### 8.5.2.1 Detection Methods

Various microsystems with defined areas for the capture and detection of target pathogen RNA and DNA sequences in materials such as PDMS,[76] poly(methyl methacrylate) (PMMA),[77] and glass, using different kinds of geometries, mass production facilities, and detection techniques, have been developed.

- In *electrochemical detection*, the use of a microchannel with gold microelectrodes integrated to detect DNA fragments through amperometric method has been reported.[78]
- *Optical detection*. In most cases, fluorescence dyes are coupled to specific target probes and used as labels. For example, using superparamagnetic beads and dye liposomes it was possible to detect the hybridization of dengue virus RNA using fluorescence microscopy.[79] Nucleic acids were also extracted for DNA and RNA analysis, using microfluidic platforms following the PCR principles.[80]
- *Electrochemiluminescent sensing*. The luminol-based reaction has applications in the fields of enzymatic biosensors, immunosensors, DNA sensors, and biochips (Figure 8.8). In this kind of detection, luminescent transitions of excited molecules or atoms to a state of lower energy are

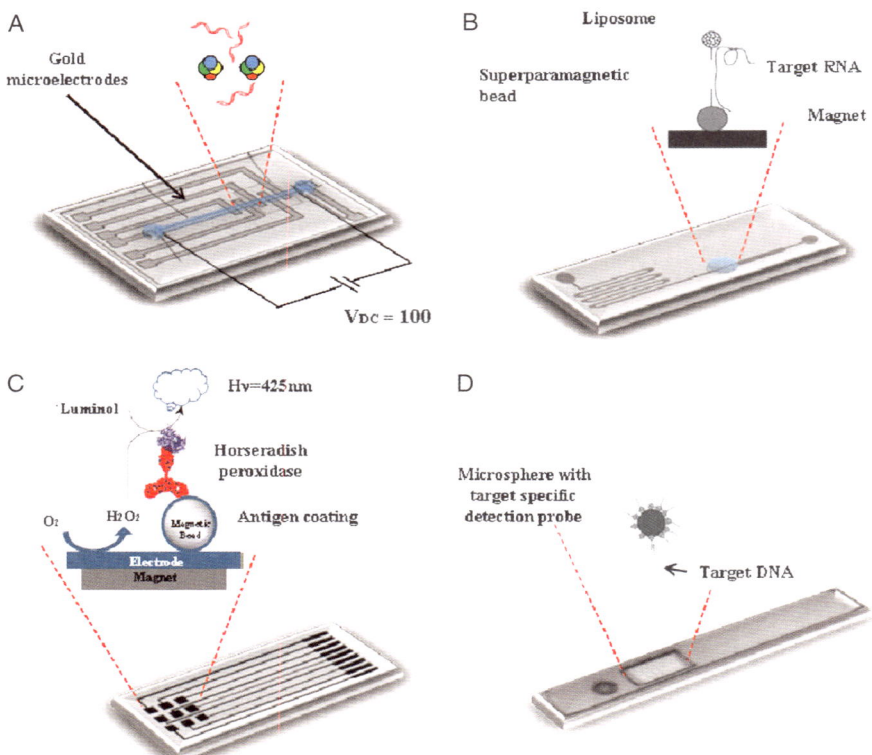

**Figure 8.8** Example of nucleic acid detection in chips. (A) Schematic diagram of a capillary electrophoresis–amperommetric detection (CE–AD) microchip. Adapted from ref. 78, with permission. (B) A DNA reporter probe is coupled to a liposome while a capture probe is coupled to a magnetic bead. When target RNA is present, the hybridization occurs. The liposome–RNA–bead complex is subsequently immobilized on a permanent magnet in the capture zone of the device. Adapted from 79 and 80, with permission. (C) In-chip electrochemiluminescent detection of glucose, lactate, or choline. Adapted from ref. 81, with permission. (D) Lateral flow device. A compact plastic housing was designed to a carry conjugate release pad and a lateral flow membrane (LFM). Adapted from ref. 82, with permission.

characterized by electromagnetic radiations dissipated as photons in the ultraviolet, visible, or near infrared (radiation related with energy involved during the excitation step).[81]

- *Chromatography*. The microfluidic concept can be combined with lateral flow principles. One example of this is a lateral flow microarray as platform for DNA detection using the hybridization-mediated target with interest for anthrax detection (Figure 8.8, D).[82]

### 8.5.2.2 Applications

NA biosensors using the concepts described above have been used to detect DNA/RNA fragments. Lab-on-a-chip platforms for toxin detection based on electrochemical transducing have been reported.[83] For example, a voltammetric DNA hybridization biosensor to detect single mismatches in ssDNA with application in toxicity studies and DNA damage caused by various compounds has been reported by Nowicka et al.[84] A microfluidic platform applied to DNA detection in environmental problems, based on PCR and a microfluidic architecture that compartmentalizes a sample into 1000 nanoliter-sized portions by centrifugation, was reported by Sundberg et al.[70] The platform has a rapid thermocycler with real-time fluorescence detection, achieving PCR cycle times of 33 s and 94% PCR efficiency. A 300-bp plasmid DNA product was amplified within the disk and analyzed in 50 min.

A PDMS microchip for capillary electrophoresis separation of DNA fragments and final amperometric detection has been reported by Joo et al.[78] The capillary is filled with 5% polyacrylamide gel, and separation of DNA fragments of different molecular weight is achieved by application of a high voltage potential across sample and waste reservoirs made at the capillary ends. The amperometric detection system involves in-channel gold microelectrodes to analyze oxidation peak for adenine residues in DNA chain, making it possible to resolve ss and dsDNA fragments with further environmental control of pathogens.

A microfluidic system for the extraction of NAs for further DNA and RNA analysis has been developed by Hui et al.[80] DNA is extracted in a multistep process by isolating and lysing white blood cells, and viral RNA is extracted directly from the submicron-sized viruses in the blood. In both cases an integrated system consisting of a mixer, valves, filters, and binders in silicon/glass microchips is used. The purity of the extracted DNA is finally demonstrated by PCR.

Lateral flow detection of DNA or RNA amplification reaction products provides a simple and cheaper method of NA detection. Indeed, a lateral flow platform may offer many significant advantages for employing NA assays. Carter et al.[82] have reported a nitrocellulose-patterning method that enables microarray feature density to attain lateral flow compatible with different substrates, making it possible to detect up to 250 fmol of target DNA using a low cost and widely available flatbed scanner, with similar sensitivity to that reported for fluorescence and chemiluminisescence detection.

Microfluidic devices have also been used for the detection of heavy metal pollutants. For example, the $Pb^{2+}$-specific DNAzyme discussed in previous sections was also tested in a microfluidic device, and was able to detect less than 1 nL of DNA.[85] The voltage-controlled microfluidic device was capable of detecting $Pb^{2+}$ concentrations at levels of 11 nM.[86] Quantitative measurements of $Pb^{2+}$ in complex samples showed this sensor to be selective and suitable to be used for the analysis of lead pollution in groundwater or drinking-water.

The previously mentioned strong interaction between $Hg^{2+}$ and bisthymine has also been applied in ionic metal detection, using capillary electrophoresis to monitor this interaction.[87] Using a dsDNA chelating fluorescence dye as marker and a polithymine oligonucleotide as probe, highly selective and sensitive detection of $Hg^{2+}$ (0.6 ppb) in pond water and in batteries is achieved after fluorescence detection.

Nie et al.[88] described the fabrication and performance of a paper-based electrochemical sensing device capable of quantifying the concentration of different analytes in aqueous solutions, including biological fluids such as urine, serum, and blood. These authors obtained measurements of heavy metal ions with higher sensitivity and limit of detection of around 1 ppb using stripping voltammetry. Following a similar methodology, Zou et al.[89] presented a heavy metal ion sensor for in situ environmental monitoring. The detection and quantification of Pb(II) and Cd(II) was performed by anodic stripping voltammetry, achieving detection limits of 8 ppb and 9.3 ppb for $Pb^{2+}$ and $Cd^{2+}$ respectively. This heavy metal detection system may be applied in future be linked to DNA detection modes, so as to increase the selectivity and sensitivity of heavy metal measurements.

## 8.6 Conclusions

DNA and nanomaterial-based sensors are proving to be interesting alternative devices for environmental applications. Nanomaterials are integrated in various ways in both electrochemical and optical detection devices. Of the various kinds of nanomaterials available, NPs such as gold NPs and quantum dots are the most reported for their use as labels, including multilabelling technologies. These applications have shown clear advantages over existing technologies. Stability, cost-efficiency, and overall multiplexing analysis seem to be the most important advantages in relation to their application in DNA-based sensors for environmental monitoring uses. Carbon nanomaterials—nanotubes, nanowires, and nanochannels—have also emerged as interesting materials to improve the performance of DNA sensors. They can lead to significant improvements in both electrochemical and optical properties, thus improving the sensitivity and detection limit for pollutant detection.

Miniaturized systems and lab-on-a-chip devices are also focusing on environmental monitoring applications. Their linkage to DNA systems is still not satisfactory, because of problems related to real sample analysis. Nevertheless, their application as screening tools preliminary to laboratory testing may open the way to the fast and in-field environmental control of specific analytes.

Although in some of DNA-related biosensing systems described in this chapter have not yet been used for environmental monitoring, their established and interesting methodologies may open the way for future applications of sensitive, simple, and cost-effective sensors of interest in this field.

Despite the achievements in the field of nanomaterials and lab-on-a-chip systems for DNA-based environmental control, some inherent problems need to be better addressed before the real advantages of these materials and technologies can be fully exploited. The complex fabrication and measuring techniques currently required are not suited to mass production, and user-friendly technologies need to be developed in to bring these devices to end-users interested in environmental monitoring.

## Acknowledgements

We acknowledge MEC (Madrid) for the projects MAT2008-03079/NAN and CSD2006-00012 "NANOBIOMED" (Consolider-Ingenio 2010) and a Juan de la Cierva scholarship (AEM), the E.U.'s support under FP7 contract number 246513 "NADINE" and the NATO Science for Peace and Security Programme's support under the project SFP 983807".

## References

1. A. Merkoçi, M. Aldavert, S. Marín and S. Alegret, *Trends Anal. Chem.*, 2005, **24**, 341.
2. A. Merkoçi, M. Pumera, X. Llopis, B. Perez, M. del Valle and S. Alegret, *Trends Anal. Chem.*, 2005, **24**, 826.
3. A. Merkoçi, *Microchim. Acta*, 2006, **152**, 157.
4. M. Pumera, A. Merkoçi and S. Alegret, *Trends Anal. Chem.*, 2006, **25**, 219.
5. C. A. Mirkin, *Inorg. Chem.*, 2000, **39**(11), 2258.
6. L. Hilliard, X. Zhao and W. Tan, *Anal. Chim. Acta*, 2002, **470**, 51.
7. D. Knopp, D. Tang and R. Niessner, *Anal. Chim. Acta*, 2009, **647**, 14.
8. P. Z. Qin, C. G. Niu, G. M. Zeng, M. Ruan, L. Tang and J. L. Gong, *Talanta*, 2009, **80**, 991.
9. A. Son, D. Dosev, M. Nichkov, Z. Ma, I. M. Kennedy, K. M. Scow and K. R. Hristov, *Anal. Biochem.*, 2007, **370**, 186.
10. G. J. Vora, C. E. Meador, G. P. Anderson and C. R. Taitt, *Mol. Cell. Probes*, 2008, **22**, 294.
11. D. J. Maxwell, J. R. Taylor and S. Nie, *J. Am. Chem. Soc.*, 2002, **124**, 9606.
12. L. A. Lyon, M. D. Musick and M. J. Natan, *Anal. Chem.*, 1998, **70**, 5177.
13. T. Endo, K. Kerman, N. Nagatani, Y. Takamura and E. Tamiya, *Anal. Chem.*, 2005, **77**, 6976.
14. A. I. Hochbaum, R. K. Chen, R. D. Delgado, W. J. Liang, E. C. Garnett, M. Najarian, A. Majumdar and P. D. Yang, *Nature*, 2008, **451**, 163.
15. P. P. Wang, B. Bai, S. Hu, J. Zhuang and X. Wang, *J. Am. Chem. Soc.*, 2009, **131**, 16953.
16. F. Lucarelli, S. Tombelli, M. Minunni, G. Marrazza and M. Mascini, *Anal. Chim. Acta*, 2008, **609**, 139.
17. S. S. Babkina and N. A. Ulakhovich, *Anal. Chem.*, 2005, **77**, 5678.
18. A. de la Escosura-Muñiz, A. Ambrosi and A. Merkoçi, *Trends Anal. Chem.*, 2008, **27**, 568.
19. J. Wang, D. Xu, A. N. Kawde and R. Polsky, *Anal. Chem.*, 2001, **73**, 5576.

20. L. Authier, C. Grossiord, P. Brossier and B. Limoges, *Anal. Chem.*, 2001, **73**, 4450.
21. J. Wang, D. Xu and R. Polsky, *J. Am. Chem. Soc.*, 2002, **124**, 4208.
22. M. T. Castañeda, A. Merkoçi, M. Pumera and S. Alegret, *Biosens. Bioelectron.*, 2007, **22**, 1961.
23. M. Pumera, M. T. Castañeda, M. I. Pividori, R. Eritja, A. Merkoçi and S. Alegret, *Langmuir*, 2005, **21**, 9625.
24. H. Cai, Y. Xu, N. Zhu, P. He and Y. Fang, *Analyst*, 2002, **127**, 803.
25. S. Marín and A. Merkoçi, *Nanotechnology*, 2009, **20**, 055101.
26. J. Wang, G. Liu and A. Merkoçi, *Anal. Chim. Acta*, 2003, **482**, 149.
27. T. M. H. Lee, H. Cai and I. M. Hsing, *Analyst*, 2005, **130**, 364.
28. J. Wang, G. Liu and A. Merkoçi, *J. Am. Chem. Soc.*, 2003, **125**, 3214.
29. X. H. Fu, *Bioprocess Biosyst. Eng*, 2008, **31**, 69.
30. J. Wang, A. N. Kawde and M. Musameh, *Analyst*, 2003, **128**, 912.
31. J. Li, H. T. Ng, A. Cassell, W. Fan, H. Chen, Q. Ye, J. Koehne, J. Han and M. Meyyappan, *Nano Lett.*, 2003, **3**, 597.
32. J. E. Koehne, H. Chen, A. M. Cassell, Q. Ye, J. Han, M. Meyyappan and J. Li, *Clin. Chem.*, 2004, **50**, 1886.
33. Y. Feng, T. Yang, W. Zhang, C. Jiang and K. Jiao, *Anal. Chim. Acta*, 2008, **616**, 144.
34. J. I. Hahm and C. M. Lieber, *Nano Lett.*, 2004, **4**, 51.
35. H. Sun, T. S. Choy, D. R. Zhu, W. C. Yam and Y. S. Fung, *Biosens. Bioelectron.*, 2009, **24**, 1405.
36. G. H. Clever, C. Kaul and T. Carell, *Angew. Chem. Int. Ed.*, 2007, **46**, 6226.
37. H. Wang, Y. Wang, J. Jin and R. Yang, *Anal. Chem.*, 2008, **80**, 9021.
38. Z. D. Liu, Y. F. Li, J. Ling and C. Zhihuang, *Environ. Sci. Technol.*, 2009, **43**, 5022.
39. H. Wei, B. Li, J. Li, S. Dong and E. Wang, *Nanotechnology*, 2008, **19**, 095501.
40. J. Liu and Y. Lu, *J. Am. Chem. Soc.*, 2003, **125**, 6642.
41. J. H. Lee, Z. Wang, J. Liu and Y. Lu, *J. Am. Chem. Soc.*, 2008, **130**, 14217.
42. Y. Lu and J. Liu, *Curr. Opin. Biotechnol.*, 2006, **17**, 580.
43. J. Liu and Y. Lu, *Anal. Chem.*, 2004, **76**, 1627.
44. S. Seetharaman, M. Zivartis, N. Sudarsan and R. R. Breaker, *Nat. Biotechnol.*, 2001, **19**, 336.
45. Z. Zhu, Y. Su, J. Li, D. Li, J. Zhang, S. Song, Y. Zhao, G. Li and C. Fan, *Anal. Chem.*, 2009, **81**, 7660.
46. Y. Zheng, C. Yang, W. Pu and J. Zhang, *Microchim. Acta*, 2009, **166**, 21.
47. Y. Peng, X. Wang, Y. Xiao, L. Feng, C. Zhao, J. Ren and X. Qu, *J. Am. Chem. Soc.*, 2009, **131**, 13813.
48. R. M. Stevens, C. V. Nguyen and M. Meyyappan, *IEEE Transactions on Nanobioscience*, 2004, **3**.
49. I. Vlassiouk, P. Takmakov and S. Smirnov, *Langmuir*, 2005, **21**, 4776.
50. I. Palchetti and M. Mascini, *Analyst*, 2008, **133**(7), 825.
51. S. E. Lyschevski, *Nano and Microelectromechanical Systems: Fundamentals of Nano and Microengineering*, CRC Press, Boca Raton, FL, 2001, p. 76.

52. S. Achiche, F. Fan and F. Bolognini, *IEEE International Symposium on Industrial Electronics*, Vigo, Spain, 2007, 2150.
53. H. J. Levinson, *Principles of Lithography*, SPIE, Bellingham, WA, 2005, **7**.
54. J. Yeom and M. A. Shannon, *Adv. Funct. Mat.*, 2010, **20**, 289.
55. J. A. Rogers and R. G. Nuzzo, *Mater. Today*, 2005, 50.
56. L. Alting, F. Kumura, H. N. Hansen and G. Bissacco, *CIRP Ann. Manuf. Technol.*, 2003, **52**(2), 635.
57. S. Mongpraneet, A. Wisitsora, R. Phatthanakun, N. Chomnawang and A. Tuantranont, *Am. Vacuum Soc.*, 2009, **27**(3), 1299.
58. J. Greener, W. Li, J. Ren, D. Voicu, V. Pakharenko, T. Tang and E. Humacheva, *Lab on a Chip*, 2010, **10**, 522.
59. D. S. Kim, S. H. Lee, C. H. Ahn, J. Y. Lee and T. H. Know, *Lab on a Chip*, 2008, **6**, 794.
60. L. P. Yeo, S. H. Ng, Z. Wang and N. F. Rooij, *Microelectro. Eng.*, 2009, **86**, 933.
61. A. Sayah, V. K. Parashar, A. Pawlowski and M. A. M. Gijs, *Sens. Actuators, A*, 2009, **125**, 84.
62. P. P. Shiu, G. K. Knopf and M. Ostojic, *Microsyst. Technol.*, 2010, **16**, 477.
63. S. Roy, *J. Phys. Appl. Phys.*, 2007, **40**, 413.
64. H. A. Stone and S. Kim, *AIChE J.*, 2001, **47**, 1250.
65. D. Wu, J. Qin and B. Lin, *J. Chromatogr.*, 2008, **1184**, 542.
66. D. Siddhartha, D. Tamal and C. Suman, *Sens. Actuators, B*, 2006, **114**(2), 957.
67. F. Grom, J. Kentsch, T. Müller, T. Schnelle and M. Stelzle, *Electrophoresis*, 2006, **27**, 1386.
68. J. D. Adams, P. Thévoz, H. Bruus and T. Soh, *Appl. Phys. Lett.*, 2009, **95**, 254103.
69. W. Zhang, S. Lin, C. Wang, J. Hu, C. Li, Z. Zhuang, Y. Zhou, R. A. Mathies and C. J. Yang, *Lab on a Chip*, 2009, **9**, 3088.
70. S. O. Sundaberg, C. T. Wittwer, C. Gao and B. K. Gale, *Anal. Chem.*, 2010, **82**, 1546.
71. V. Adam, D. Huska, J. Hubalek and R. Kizek, *Microfluid. Nanofluid.*, 2010, **8**, 329.
72. A. Agarwal, K. Buddharaju, I. K. Lao, N. Singh, N. Balasubramanian and D. L. Kwong, *Sens. Actuators, A*, 2008, **145**, 207.
73. M. Lee, K. Y. Baik, M. Noah, Y. K. Kwon, J. Lee and S. Hong, *Lab on a Chip*, 2009, **9**, 2267.
74. D. Kim, J. Park, J. Shin, P. Kim, G. Lim and S. Shoji, *Sensors and Actuators B*, 2006, **117**, 488.
75. H. Ogi, Y. Fukunishi, H. Nagai, K. Okomoto, M. Hirao and M. Nishiyamo, *Biosens. Bioelectron.*, 2009, **24**, 3148.
76. S. Kwakye and A. Baeumner, *Anal. Bioanal. Chem.*, 2003, **376**, 1067.
77. S. R. Nugen, P. J. Asiello, J. T. Connelly and A. J. Baeumner, *Biosens. Bioelectron.*, 2009, **24**, 2428.
78. G. S. Joo, S. K. Jha and Y. S. Kim, *Curr. Appl. Phys.*, 2009, **9**, 222.
79. C. Zhang and D. Xing, *Nucleic Acids Res.*, 2007, **35**, 4223.

80. W. C. Hui, L. Yobas, V. D. Samper, C. K. Heng, S. Liw, H. Ji, Y. Chen, J. Cong, J. Li and T. M. Lim, *Sens. Actuators, A*, 2007, **133**, 335.
81. C. A. Marquette and L. J. Blum, *Anal. Bioanal. Chem.*, 2008, **390**, 155.
82. D. J. Carter and R. B. Cary, *Nucleic Acid Res.*, 2007, **1**(11), 1.
83. I. Palchetti and M. Mascini, *Anal. Bioanal. Chem.*, 2008, **391**, 455.
84. A. M. Nowicka, A. Kowalczyk, Z. Stojek and M. Hepel, *Biophys. Chem.*, 2010, **146**, 42.
85. K. A. Shaikh, K. S. Ryu, E. D. Goluch, J. Nam, J. Liu, C. S. Thaxton, T. N. Chiesl, A. E. Barron, Y. Lu and C. A. Mirkin, *Proc. Natl. Acad. Sci. U. S. A.*, 2005, **102**, 9745.
86. I. H. Chang, J. Tulock, J. Liu, W. S. Kim, D. Cannon, J. R. Y. Lu, P. W. Bohn, J. Sweedler and D. Cropek, *Environ. Sci. Technol.*, 2005, **39**, 3756.
87. C. Chiang, C. Huang, C. Liu and H. Chang, *Anal. Chem.*, 2008, **80**, 3716.
88. Z. Nie, C. A. Nijhuis, J. Gong, X. Chen, A. Kumachev, A. Martínez, M. Narovlyasky and G. M. Whitesides, *Lab on a Chip*, 2010, **10**, 477.
89. Z. Zou, A. Jang, E. MacKnight, P. Wu, J. Do, P. L. Bishop and C. H. Ahn, *Sens. Actuators, B*, 2008, **134**, 18.

CHAPTER 9
# *Conclusions and Criticisms*

ILARIA PALCHETTI AND MARCO MASCINI

Dipartimento di Chimica, Università degli Studi di Firenze, 50019, Sesto Fiorentino (Fi), Italy

The chemical analysis of environmental pollutants is predominantly performed by implementing analytical techniques that have the sensitivity, reproducibility, and reliability to determine subnanomolar concentrations of analytes in complex matrices. Gas or liquid chromatography coupled to mass spectrometry (CG-MS or HPLC-MS) is necessary to detect organic pollutants, and atomic spectroscopy is essential for trace and ultratrace analysis of metals. The measurement of compounds in environmental media, however, involves several steps in addition to analysis by instrumental techniques, such as sampling, transport to the laboratory, extraction, and clean-up. Each of these steps requires time, effort, and expense. Biosensors and bioanalytical methods appear well suited to complement standard analytical methods for a number of environmental monitoring applications, since they have the advantage of being simple, rapid, cost-effective, and field-portable screening methods. In particular, biosensors are considered the best choice when real-time, continuous monitoring is required.

Nucleic acids (NAs) are undoubtedly excellent molecules for the development of smart and innovative biosensors for environmental monitoring, as summarized in this book. Genosensors offer considerable promise for obtaining sequence-specific information in a faster, simpler, and cheaper manner than the traditional hybridization assay. Many types of DNA biosensors have been proposed for the detection of DNA damage and genotoxicity. Moreover, although they have appeared in the literature only recently, functional NA molecules such as aptamers, DNAzymes, and aptazymes have already found applications in environmental monitoring.

Nanomaterials (carbon nanotubes, metal nanoparticles, or metal oxide nanostructures) will definitely have a great impact on development of NA-based biosensors for environmental monitoring. Nanotechnology brings new possibilities for biosensor construction and for developing novel bioassays. Nanoscale materials have been used to promote reaction, to impose nano-barcodes for biomaterials, and to amplify the signal of biorecognition events. Moreover, in the near future, we are likely to see the incorporation of individual devices into more complex multifaceted systems, also called micro total analysis systems (μTAS) or "Lab-on-a-chip" (LOC)—a term used for devices that integrate (multiple) laboratory functions on a single chip of only millimetres to a few square centimetres in size and that are capable of handling extremely small fluid volumes, down to less than picoliters.[1]

The integration of nanotechnology, microfluidics, and bioanalytical systems clearly represents one of the future directions of all biosensor research. When these different areas are combined the possibilities seem endless, and there are high hopes of solving many current problems such as sample preparation, real portability, single-molecule detection, analytical speed, and reliability. Many obstacles will have to be overcome in order to fulfill all of these hopes; however, some exciting examples, typically of subsystems rather than complete bioanalytical sensors, can already be found in the current literature. Miniaturization of polymerase chain reaction (PCR) devices, for instance, offers several advantages such as short assay time, low reagent consumption, and rapid heating/cooling rates, as well as great potential for integrating multiple processing modules to reduce size and power consumption. This will undoubtedly benefit areas such as genosensor assays, since multiple processes, including sample collection and pretreatment, DNA extraction, amplification, hybridization, and detection can be performed on a single, self-contained microfluidic platform. Such miniaturization of the analytical instrumentation will enable transportation of the laboratory to the sample source, as required for point-of-care testing. In conclusion, in the near future, NA sensing and biosensor technology itself will undoubtedly benefit from nanotechnology and μTAS technology.

There are, however, some issues that must be considered when NA biosensors are viewed in the context of environmental monitoring requirements. The most important of these is that biosensors and screening assays have been described for only a limited number of pollutants, whereas it is estimated that there are about 80 000 chemicals in current use and more than 1000 new products reach the market every year.[2] Moreover, environmental matrices (water, soil, air) are so heterogeneous and complex that biosensor response in these matrices may be completely different from what is expected in buffer or almost clean solution, and may also differ from one matrix to another. Thus rigorous programs of analysis of real samples should be performed to check the performance of biosensors. These issues, together with the variability of data quality requirements among environmental programs are important obstacles to the commercialization of NA biosensors and biosensors in general. Nevertheless, as already mentioned, continued advances in the areas of nanomaterials and micro- and nanoenginering could broaden the potential market and allow these techniques to be adopted for routine use.

## References

1. C. Zhang and D. Xing, *Nucleic Acids Res.*, 2007, **35**, 4223–7.
2. M. Farre and D. Barceló, (Eds.), *Biosensors for the Environmental Monitoring of Aquatic Systems. Bioanalytical and Chemical Methods for Endocrine Disruptors. Handbook of Environmental Chemistry, Vol. 5: Water Pollution, Part 5J*, Springer, Berlin, 2009, p. 115.

# Subject Index

abrin 73
acoustic detection 5, 46, 49, 105, 106–7, 156–7
adenine oxidation 49, 103, 126, 131–2, 153, 159
adsorption immobilization 38, 40
affinity columns 61, 64–6, 68, 73
affinity constants 63
affinity immobilization 37, 40
  *see also* biotin; streptavidin
AFM (atomic force microscopy) 26, 51, 124
algae 54, 56
allosteric DNAzymes (aptazymes) 12–13, 87, 93–4, 150
ALP (alkaline phosphatase) 27, 50–1, 53
Ames test 100, 102, 111–12, 116
amplification *see* NASBA; pretreatments
analytical techniques and biosensors 1–2, 165
anions and catalytic nucleic acids 95
anthrax 66, 73, 76, 143, 158
antibiotics 71–2
antibody-based biosensors 7–9
  aptamer advantages over 11, 68, 72
antioxidants 131
aptamer-based biosensors 67–78, 150
  pathogens 75–7
  pharmaceuticals 70–2
  pollutants 68–70
  toxins 72–5

aptamer magnetic electrochemiluminescence assay (AM-ECL) 76
aptamers 10–13, 28, 61–3
aptazymes 12–13, 87, 93–4, 150
aromatic amines 103, 109
asymmetric PCR 42, 54
atomic absorption spectroscopy (AAS) 83
AuNPs (gold nanoparticles) 69–70, 89–94, 142–4, 146, 148–53
  DNAzyme conjugates 91–3
avidin 53, 76
  *see also* biotin; streptavidin

bacteria, pathogenic *see* pathogenic organisms
bacterial assays 102, *105–6*, *112*, 127
  Ames test 100, 102, 111–12, 116
benzo[a]pyrene 102, 108, 110, 115
BIAcore instrument 3
bio-barcodes 93
bioanalytical systems
  biosensors distinguished from 4, 101
  DNA damage detection 104–6
biological oxygen demand (BOD) 2
bioluminescence detection 105–6
biorecognition elements 6–14
biosecurity 72
biosensor arrays
  bacterial 114–15
  DNA microarrays 35, 40, 48, 88
biosensors
  antibody-based biosensors 7–9

## Subject Index

aptamer-based 67–77
biorecognition elements 6–14
capture probe
   immobilization 37–40
definition, classification and
   history 3–6, 101
DNA biosensors 25–7
environmental monitoring
   applications 1–3
LNA biosensors 29–30
nanomaterials as 148–53
overview of NA-based
   technologies 21–30
PNA biosensors 28–9
RNA biosensors 27–8
single use biosensors 4
transduction principles 4–6
whole cell 110–16
*see also* DNA damage detection;
   genosensors
biotin 77, 124, 148
   biotin-avidin interactions 39, *71*, 124
   biotin-neutravidin
     interactions 148
   biotin-streptavidin interactions
     (*see* streptavidin)
BLAST (basic local alignment search
   tool) 36
BOD (biological oxygen demand) 2
botulinum toxins 74–5
'break lights' assay 110
bridge nucleic acids *see* LNA

cadmium ($Cd^{2+}$) 103, 132, 160
CAMB (catalytic and molecular
   beacon method) 86–7, 94
cantilevers *see* micromechanical
   transduction
capillary electrophoresis (CE) 84, 159
   CE-SELEX 66, *67*, 72–3
capture probes
   genosensor design and 36–7
   probe immobilization 37–40, 54
carbon-based electrodes 123
   carbon paste electrodes
     (CPE) 22–4, 123, 126, 129

glassy carbon electrodes (GCE) 23,
   123, 126, 129, 148, 153
pyrolytic graphite electrodes
   (PGE) 22–4, 123, 126, 128
screen-printed carbon electrodes
   (SPCE) 22–4, 123, 126, 129
carbon nanotubes 147, 153–4
   multiwalled carbon nanotubes
     (MWNTs) 28, 130, 153
   single-walled nanotubes
     (SWNT) 73, 75–6
catalytic beacon (and molecular
   beacon, CAMB) method 86–7, 94
catalytic nucleic acids
   aptazyme use with 93–5
   conversion to biosensors 85–93
   trace contaminant detection 83–4
   *in vitro* selection 84–5
   *see also* DNAzymes
cathodic inidicators 129
chemiluminescence 5, 27, 44
   electrochemiluminescence
     (ECL) 69–70, 76–7, 132
chemisorption 38
cholinesterase 6
cobalt ($Co^{2+}$) 85
cocaine 68–70, 91
colorimetry 44, 52, 94–5, 115,
   149–51
   aptamer-based biosensors 69–70
   DNAzymes 89–91
   nanoparticles and 149–51
comet assay 100, 107, 122, 127
commercial biosensors 3, 14
commercialization 2–3, 118, 161, 167
copper ($Cu^{2+}$) sensing 85, 87–8,
   91, 94
core–shell nanoparticles 129–30,
   143–4, 146
covalent probe immobilization 37–9
CPE (carbon paste electrodes) 22–4,
   123, 126, 129
cyclic voltammetry (CV) 26, 71–2, 88,
   130, 148
cytochrome C 131
cytochrome P450 102, 110, 132

damage *see* DNA damage detection
daunomycin 26, 128–9
denaturing samples 47, 125
deoxyribozymes *see* DNAzymes
differential pulse voltammetry (DPV) 26, 29, 148
dipstick tests 91–3
direct PCR 42, 54
disposable electrochemical printed (DEP) chips 57
DNA aptamers 63, 76
DNA biosensors 25–7
  *see also* genosensors
DNA damage
  intercalation 27, 50, 73, 102–9, 129–30
  radiation-induced 108–9
  strand breaks (SB) 102–6, 118, 121–2, 125–9, 132–3
  types of damage 102–3, 108, 121–2, 128
DNA damage detection 99–107, 110–16, 122–3
  biosensor detection 12, 101–2
  electrochemical transducers 123–33
  general assays 100
  optical transducers 107–16
  piezoelectric transducers 133–4
  recent advances 104–7, 116–18
DNA (deoxyribonucleic acid)
  hydrolysis resistance 11
  RNA distinguished from 17–18
DNA hybridization biosensors *see* genosensors
DNA imaging 48
DNA surrogates 101–3, *104–6*, 107, 109–10, 117–18
DNAzymes 11–13, 82–3
  allosteric (aptazymes) 12–13, 87, 93–4, 150
  AuNP conjugates 91–3
  fluorescence detection 86–9
  metal ion sensing by 85–93
  transduction methods 86–93
  *see also* catalytic nucleic acids

dyes
  affinity columns 61, 65
  colorimetry 94
  intercalating 103, 107–9
  methylene blue (MB) 26, 29–30, 93, 129–30, 153–4
  nanoparticles and 143, 146, 149
  PicoGreen (PG) 107–9
  Raman dyes 44
  thiazole oprange (TO) 107, 109
  *see also* fluorescence detection

*E. coli see Escherichia*
ECL (electrochemiluminescence) 69–70, 76–7, 93, *105*, 132, 157
  AM-ECL 76
EIS (electrochemical impedance spectroscopy) 26, 29, 49, 124, 131, 148
electrochemical quartz crystal nanobalance (EQCN) 134
electrochemical sensors 5, *22–4*, 25–6, 29, 45, 49, 160
  aptamer-based biosensors 74–5
  DNA damage detection 122–33
  DNA microarrays 35, 40
  DNAzymes 93
  environmental monitoring 160
  LNA-based 29
  nanomaterial-based 146–8, 153–4, 157
  transducer types 5, 22–4, 25–6, 45, 49
electrode design 123
electroluminescence 102
electropolymerization 39, 40, *41*
ELISA 2
endocrine-disrupting chemicals (EDCs) 70–2
engineered protein scaffolds 10
environmental pollution, forms of 52
Environmental Protection Agency, US (EPA) 83, 86, 95
enzyme-based biosensors 6, 11, 82

enzyme electrodes/enzyme
   transducers 3
EPA (US Environmental Protection
   Agency) 83, 86, 95
*Escherichia coli*
   DNA damage detection 113–15,
      128
   environental sensing 53–4, 56–7,
      143, 148
   RNA biosensors 27–8, 75–6
estrogens 71–2
explosives 77

Fab (fragment antigen binding) 8–9
false-negative results 43
false-positive results 43, 56
Fc (fragment crystallizable) 8–9
fecal contamination 54–6
Fenton reaction 104–5, 110, 126,
   129–31
ferrocenes 3, 29, 45, 128, 132–3
FETs (field-effect transistors) 49, 157
fiber-optic biosensors 27, 77, 109,
   116, 132
fluorescence detection 44
   aptamer-based detection 68–9,
      72–3, 77
   catalytic nucleic acids 86–9
   DNA damage 105–6
   intercalating dyes 107, 129
   lab on a chip devices 157, 160
   real-time PCR 42
   redox inidicators 129
fluorescence spectroscopy 4–5
food contaminants/pathogens 42,
   52–7, 75
foods, genetically modified 57
fragmentation of samples 43
free radicals 102–3, 122–3, 125–6
   hydroxyl radicals 110, 114, 129–31
FRET (Förster/fluorescence
   resonance energy transfer) 44, 46,
   68–9, 110

GCE (glassy carbon electrodes) 23,
   123, 126, 129, 148, 153

GEC (graphite–epoxy composite
   magneto electrode) 53–4
gel electrophoresis 41–2, 84
GEMs (genetically engineered
   bacteria) 7
gene chips *see* biosensor arrays
genetic engineering 6, 102, 117
genetically modified foods 57
genosensors 10
   capture probe design 36–7
   environmental monitoring 51–7
   principles and development
      of 35–7
   RNA based 27–8
   sample treatment and
      hybridization 40–7
   transducer design 25–7, *44–6*,
      47–51
genotoxicity screening 100–1, 103–6,
   110–18, 132, 148
   *see also* DNA damage detection
genotoxins 100–2, 108–16, 122–3,
   126, 130
GFP (green fluorescent
   protein) 113–14, 116
giant magnetoresistive effect
   (GMR) 6
glucose oxidase 3, 51
glycine, *N*-(2-aminoethyl)- 19
gold nanoparticles (AuNP) 69–70,
   89–94, 142–4, 146, 148–53
graphite–epoxy composite magneto
   electrode (m-GEC) 53–4
guanine oxidation 51, 110
   carbon electrodes 125–8, 131,
      153
   voltammetric detection 26,
      49, 110

hairpin capture probes 36–7
heavy metals
   aptamer-based detection 68–70
   microfluidic detection 159–60
   nanomaterial-based
      detection 148–53
history of biosensors 3–6

HMDE (hanging mercury drop electrode) 22-4, 126
HRP (horseradish peroxidase) 26, 50-1, 54, 56, 74-5, 147
HRP mimicking DNAzyme 94
human cell biosensors 116
hybridization biosensors *see* genosensors
hybridomas 8
hydrolysis of nucleic acids 11, 83, 84, 102-3, 121, 125
hydroxyl radicals 110, 114, 129-31

IgG (immunoglobulin G) 7-8
imaging hybridization detection 50-1
immobilization
 affinity immobilization 37, 40
 electrochemical biosensors 124, 131
 terminal capping 63
immunoglobulin G (IgG) 7-8
immunosensors 7-9
 aptamer advantages over 11, 68, 72
impedimetric detection 49-50, 124, 131
 EIS 26, 29, 49, 124, 131, 148
 PQCI 133
*in vitro* selection 84-5
indicatorless detection 125
inductively coupled plasma (ICP) 86
inductively coupled plasma mass spectrometry (ICP-MS) 83
inorganic nanocrystals 147
intercalation 27, 50, 73, 102-9, 129-30
interferometry 44
IQ (2-amino-3- methylimidazo[4,5-f]quinoline) *105-6*, 116
ITO (indium tin-oxide) electrodes 123, 125, 146
IUPAC (International Union of Pure and Applied Chemistry) 4, 122

Kelvin probe force microscopy (KPFM) 50-1

'lab on a chip' (LOC) technology 2, 101, 155-60
label-based detection methods 50, 91, 128-9
label-free detection methods 48-50, 87, 91, 125-8, 148
lateral flow devices 91-2, 158-9
lead ($Pb^{2+}$)
 DNA damage 103, 132
 sensing by DNAzymes 85-94, 150, 159-60
*Legionella pneumophila* 53
leukemia 29-30
ligation 67, 83, 91
linear capture probes 36
LNA biosensors 29-30, 36
LNA (locked nucleic acid) 21, 24
loop-mediated isothermal amplicon (LAMP) 57
LSPR (localized surface plasmon resonance) 144-5
Lux proteins 113-15

magnetic assays 26-7, 30, 53
magnetic beads 40, 54-5, 65, 76
mass sensitive devices 23-4
 piezoelectric biosensors 46, 133-4, 148
melting points 19, 21
meningitis 26
mercury electrodes 24, 26, 123, 125-6
 hanging mercury drop electrode (HMDE) 22-4, 126
 solid amalgam electrodes (SAE) 123, 126
mercury ($Hg^{2+}$)
 bisthymine reaction 149-50, 152-3, 160
 sensing by DNAzymes 85, 87
mercury-specific oligonucleotide (MSO) 152-3
metal ion sensing 85-93

Subject Index

metal nanoparticles 50, 146
  gold nanoparticles
    (AuNPs) 69–70, 89–94, 142–4,
    146, 148–53
  silver nanoparticles
    (AgNPs) 146–8
metals
  transition metals 102–3, 150
  see also heavy metals
methylene blue (MB) 26, 29–30, 93,
  129–30, 153–4
micro total analysis systems
  (µTAS) 155
microelectromechanical system
  (MEMS) 155
microfluidic devices 88–9, 156–9,
  166
microgravimetric transduction
  49, 133
  see also mass sensitive devices
micromechanical transduction 5, 46
microRNAs (miRNAs) 18
microtechnology 155
MIPs (molecularly imprinted
  polymers) 13–14
mitomycin C 102, 114–15, 132
MMS (methyl methanesulfonate)
  102, *105–6*, 108, 128
MNNG (1-methyl-1-nitroso-*N*-
  methylguanidine) *106*, 114–15
molecular beacon method 87,
  132–3
molecularly imprinted polymers
  (MIPs) 13–14
monoclonal antibodies 8
multiplex PCR 42
multiwalled carbon nanotubes
  (MWNTs) 28, 130, 153
mutagenicity 12, 111, 122, 148

nanocapillary membranes
  (NCAMs) 88
nanofluidic devices 88–9
nanomaterials
  as biosensors 148–53, 166
  carbon nanotubes 28, 141, 157

core-shell nanoparticles 129–30,
  143–4, 146
DNA damage screening 117, 124
as DNA sensors 50, 141–8
magnetic nanobeads 40
multiwalled nanotubes
  (MWNTs) 28, 130, 153
nanopore and nanochanel
  sensing 153–4
one-dimensional (nanowires,
  NWs) 144, 148, 157
quantum dots (QD) 77, 146–7
silica nanoparticles 143–4
single-walled nanotubes
  (SWNT) 73, 75–6
NASBA (nucleic-acid sequence
  based amplification) 42–3, 52,
  56–7
nickel ($Ni^{2+}$) 103, 132
nitrocellulose filtration 64, 66, 72
nomenclature recommendations 4
noncovalent attachment 13–14,
  102–3, 121, 124, 129
nuclease resistance 63
nucleic acid enzymes see DNAzymes
nucleic acids (NA)
  natural, described and
    classified 17–18
  rival biorecognition elements
    and 10–14
  synthetic, described and
    classified 18–21
  see also catalytic nucleic acids;
    DNA; RNA

oligodeoxyribonucleotides
  (ODNs) 25, 51, 128, 133
oligonucleotides
  capture probes 36–40
  as DNA surrogates 103–7, 110,
    117
  mercury-specific oligonucleotide
    (MSO) 152–3
optical detection
  DNA damage 107–16
  lab-on-a-chip 157

optical detection (*continued*)
  nanomaterial-based
    sensing 149–50
  nanopaticle labels and 142–6
  types of 4–5, *23–4*, 27, 44–5

paraquat (methyl viologen) 114
pathogenic organisms
  aptamer-based detection 75–7
  nanomaterial-based
    detection 142–8, 157–60
  probe design 36
  RNA biosensors for 27–8
  species identification 52, 54–7
  *see also* viruses
PDMS
  (poly(dimethylsiloxane) 155–7, 159
pesticides 6
PGE (pyrolytic graphite
  electrodes) 22–4, 123, 126, 128
phage display 9–10
pharmaceuticals and personal care
  products (PPCPs) 67, 70
Pharmacia Biosensor 3
photochemical SELEX 67
photolithography 155
photoluminescence 27
physicochemical transducers 4, 48, 122
  *see also* transduction principles
piezoelectric biosensors 46, 133–4, 148
  QCM 5, 28, 46, 49, 53, 132–3
piezoelectric quartz crystal impedance
  (PQCI) 133, 148
PNA (peptide nucleic acid) 18–21, 24, 144
  PNA biosensors 28–9, 36
pollutants
  aptamer-based detection 68–70
  endocrine-disrupting chemicals
    (EDCs) 70–2
  phenols 153
  unmanageable numbers of 99–100, 166
polyclonal antibodies 8

polycyclic aromatic hydrocarbons
  (PAHs) 108, 122, 131
  benzo[a]pyrene 102, 108, 110, 115
polymer-based DNA sensors 26
polymer probe immobilization 37
polymerase chain reaction
  (PCR) 41–2, 52–3, 84, 144
  aptamers and 63–4
PQC[I] (piezoelectric quartz crystal
  [impedance]) biosensors 133, 148
pretreatments, biosensors 102, 123
  genosensor preamplification 40–3
  LOC devices 156–7
probes, genosensors *see* capture
  probes; signaling probes
proteins scaffolds 10
pseudoknot capture probes 37
pyrenes 26, 73, 108

QCM (quartz crystal microbalance)
  transducers 5, 28, 46, 49, 53, 132–3
quantum dots (QD) 77, 146–7
quinones 25, 29

reactive oxygen species (ROS) 114, 122, 126, 129
reagentless sensing 48
real-time PCR 42
recombinant antibodies 8–10
recombinant DNA technology 6–7
redox mediators 127
redox probes 50–1, 129–31
refractive index-based transducers 49
restriction enzymes 108–9, 133
reverse transcriptase PCR
  (RT-PCR) 10, 42
ribosome display 10
ribozymes 11, 82, 94
  *see also* DNAzymes
ricin 72
RNA aptamers 63, 65–6
RNA biosensors 27–8
RNA (ribonucleic acid)
  DNA distinguished from 17–18
  DNAzyme cleavage 12
'RNA world' 82

Subject Index

SAE (solid amalgam electrodes) 123, 126
sandwich hybridization assay 26, 47–8, 53, 56, 146–7
SAW (surface acoustic wave) sensors 46, 49
scanning probe microscopy 50, 52–3, 153–4
SECM (scanning electrochemical microscopy) 40, *41*, 51
SELEX (systematic evolution of ligands by exponential enrichment) 10–11, 61, *62*, 63–7
  CE-SELEX 66, *67*, 72–3
  SPR-SELEX 64–5, 76
  variants of 64–7
self-assembly immobilization 37
SERS (surface-enhanced Raman scattering spectroscopy) 44, 46
SH-SAW (SH-SAW, shear horizontal surface acoustic wave) devices 46
signal-on and signal-off techniques 131
signaling probes 48
silica nanoparticles 143–4
silver nanoparticles (AgNPs) 146–8
single-base mismatches 29, 117
single use sensors 4, 26
single-walled nanotubes (SWNT) 73, 75–6
soft lithography 155–6
solid-phase extraction (SPE) 14
solid-supported DNAzyme sensors 88
solution-based biosensors 86–7, 130
solution-based SELEX 66
SOS reponse 111, 113–15
SPCE (screen-printed carbon electrodes) 22–4, 123, 126, 129
spin-value GMR 6
SPR *see* surface plasmon resonance
square wave voltammetry (SWV) 25, 29, 71–2, 110, 131–2
steric effects 25, 40, 48, 65–6, 130
storage of bacterial biosensors 115–16
storage of NAs 83, 88, 91
strand breaks (SB) 102–6, 118, 121–2, 125–7, 129
  single-strand breaks (SSB) 125, 128, 132–3
streptavidin-ALP 53
streptavidin-biotin 37, 40, 48, 71–2, 74–6, 91–2
  DNA damage detection 128
  environmental monitoring 51, 53–4
streptavidin-HRP 26
stringency 11, 47, 64, 85
structured capture probes 36–7
styrene 110, 132
styrene oxide (SO) 102, 108, 130, 132
subtle damage *see* DNA damage
surface plasmon effect 89
surface plasmon resonance (SPR) 27, *45*, 48–9, 51, 57, 72
  localized SPR (LSPR) 144–5
  nanomaterial-based sensors 142, 150
  SPR-SELEX 64–5, 76
surrogate DNA 101–3, *104–6*, 107, 109–10, 117–18
synthetic aptamer methods 11
synthetic nucleic acids 18–21
  *see also* LNA; oligonucleotides; PNA
systematic evolution of ligands by exponential enrichment *see* SELEX

'tailored' SELEX 67
tetracyclines 71
thermal stability 19, 29
TNT (trinitrotoluene) 77
'toggle' SELEX 66–7
toxicity
  use of biosensors 2
  use of genomic DNA 12
  use of screening assays 100
  *see also* genotoxins
toxins 66–7, 72–5, 159
  *see also* anthrax

transduction principles
　biosensor classification by 4–6
　DNA damage sensors 107–10, 122
　DNAzymes 86–93
　genosensors *44–6*, 47–51
　nanoparticle effects 147–8
transition metals 102–3, 150
tyrosinase 6

uranium ($UO_2^{2+}$) sensing 85–6, 88–91, 150–1
US Environmental Protection Agency (EPA) 83, 86, 95

viologens 114, 128
viruses 25–6, 42, 56, 159
　dengue 28, 157
　influenza 65, 76

water pollution
　algae 54, 56
　fecal contamination 54–6
　toxicity tests 127
whole cell biosensors 110–16
whole-cell systems 6–7

yeasts 7, 10
　DNA damage detection 102, *105–6*, 113, 116
　as DNA surrogates 102, *105–6*
　genetically modified 7, 102
　RNA biosenors 28

zinc ($Zn^{2+}$) 68–70, 85